The World That Is

by

Alan Weatherley

DREAMCATCHER PUBLISHING
Saint John • New Brunswick • Canada

Canadian Cataloguing in Publication Data

Weatherley, A. H. - 1928

The World That Is

ISBN - 1-894372-19-0
              I. Title.
        PS8595.E165W67 2002   C813'.6   C2002-902898-1
        PR9199.4.W42W67 2002

Editor: Yvonne Wilson

Typesetter: Chas Goguen

Cover Photo: Alan Weatherley

Cover Design: Dawn Drew, INK Graphic Design Services Corp.

Printed and bound in Canada

DREAMCATCHER PUBLISHING INC.
1 Market Square
Suite 306 Dockside
Saint John, New Brunswick, Canada E2L 4Z6

"I believe that a conscious affinity with Nature forms the shield of Perseus through which man can affront the Gorgon of his fate . . ."

- From *The Unquiet Grave* by Cyril Connolly

To Dilys
with best wishes and love

Alan

SUMMARY

The World That Is is a fictional account of an Australian biologist becoming an ecological conservationist at a time when the foundations of this subject, as we now know it, were first being established. Much of his research is carried out in Australia and North America, though his work also carries him to Europe and the Far East. Intertwined with the story of his career is that of the young American woman with whom he falls in love and marries as she becomes a leading film actress. The World That Is deals with the struggle of two people in love to pursue diverse, demanding careers, while remaining true to their professional ideals and to each other.

Sincere thanks to Yvonne Wilson for her patient, detailed and wise editorial advice – nearly all of which I have taken.

For Bobbie, whose encouragement never flags.

Other Works by Alan Weatherley

Fiction:

Short stories in The Toronto Star short fiction contests, and in The New Brunswick Reader

Non-Fiction:

Articles in The New Brunswick Reader

Scientific Books:

Australian Inland Waters and their Fauna: Eleven Studies, A.H. Weatherley, Editor, 1967, Australian National University Press.

Growth and Ecology of Fish Populations, by A.H. Weatherley, 1972, Academic Press.

The Biology of Fish Growth, by A.H. Weatherley & H.S. Gill, 1987, Academic Press.

Foreword

P. S. Lake
Professor, School of Biological Sciences
Monash University
Clayton, Victoria
Australia

This bold and brisk novel is a trans-Pacific tale mainly set in USA and Australia. It basically recounts through the eyes of a young Australian male biologist the trials and tribulations of the quest for the love of his life and for a fruitful career in ecology. The novel has a strong sense of place with evocative descriptions of such scattered locations as the Tasmanian high country, the mining towns of Virginia, the coral cays of The Great Barrier Reef, the high plains of Wyoming, the "tambaks" of Indonesia. Not surprisingly, given the background of the author, the novel presents a very authentic account of how many ecologists practice their profession.

Set in the sixties, the novel addresses the difficult dilemma of how to reconcile the ambitions of people in a loving partnership who wish to pursue their own independent careers. The young and up-and-coming biologist, Bill Logan, and a vibrant actress of great talent, Betty McMurtry, fall in love and marry. But their unconventional union is marred by the difficulties of maintaining their strong partnership while, in pursuing their own careers, they are forced to live apart for long periods.

The grand theme of the book is concerned with the development of ecology, from being "scientific natural history" to "the subversive science". Whilst a few novelists have addressed conservation strugg-les and environmentalism, "The World That Is" appears to be unique in being focused on the intellectual tenets and problems of ecology as a science. As depicted in the book, ecology is subversive as it does not follow the reductionistic form of analysis of the traditional sciences; and as ecologists investigate the animals and plants of this world in their communities and ecosystems, they unearth knowledge and truths greatly at odds with society's demand for growth -

often wasteful and unsustainable growth.

Conventionally, scientists are firmly objective in their work and have been loathe to indulge in public debate and politics. This objectivity and non-participation in public controversy has been the norm and expected stance in academic circles. As the novel clearly portrays, ecologists in studying the biosphere and confronting environmental problems follow a difficult if not tortuous path in trying to remain objective and have academic respect and in trying to provide knowledge to the public on environmental problems.

At the time of the novel, after a long period of slow and spasmodic development with only a small number of adherents, ecology began to grow rapidly in strength and in its capacity to resolve applied problems. We see this transformation through Bill Logan's ecological projects that increasingly become less theoretical and more applicable to conservation problems. For as ecology began to mature in the sixties so did the conservation movement - a movement that rapidly developed into a political force to be reckoned with even in that time of unchained developmentalism. From Bill Logan's dealing with particular ecological problems at particular locations, we see the revelation develop that society is creating not just many small problems but large - world-scale - environmental problems. Thus the novel unveils the current predicament - for example, the National Academy of Science has just announced that humans are utilizing the natural resources of the world unsustainably - more rapidly than they can be replenished. The quest of the novel's biologist to try and reconcile conservation and ecological sustainability with development and growth which is now more poignant and yet more unresolvable.

So "The World That Is" is a bold novel for our turbulent times. It addresses difficult and crucial issues and in its straightforward way offers both hope and warnings.

Chapter 1

I think I began to become a biologist one bright afternoon when I was ten. I was with my mother on the ferry wharf at Watson's Bay, and we were staring up in wonder at a hammerhead shark. It was 1938, and my uncle had caught the shark that morning in an ocean fishing contest that was part of the Australian sesquicentennial celebrations. The shark, which hung on a rusty iron hook suspended from a wooden scaffold, looked monstrous and menacing – but magnificent, even in death – against the brilliant sky. And all around us the seagulls wailed, and we could hear the lions roar across Sydney Harbour at Taronga Park Zoo. And everywhere creatures seemed to swim and rise and sail and sing in the blue glare of summer under the white actinic sun.

*   *   *

Yet often since that time I have thought I was lucky to have got through four undergraduate years of biology without being totally put off the subject.

The level of teaching in biology – particularly zoology – was largely uninspired, the material dull and old-fashioned, and we used British textbooks and laboratory examples. And as we worked with these inappropriate tools, the marvellous creatures of the Australian biological world pulsed all around us every day, no further away than a park, a stretch of shore . . . or a bushy back yard.

It wasn't till my last year as an undergraduate that I got from a lecturer something of that surge of excitement that scientists often recall as the trigger of their later careers. The odd thing is that I received it from a man I could never really like.

He was Percy Swanson, a strong, square-built Englishman from Oxford, the zoology department's most recent appointment.

Swanson was an ecologist, but with a difference. What little we had learned of ecologists led us to think of them as naturalists – essentially field workers, studying live animals and plants in a state of nature. But Percy Swanson was of a new breed, one that was not so uncommon in Britain, Europe or America, but unfamiliar to us. He experimented on tiny beetles – weevils – that he kept in glass jars, in which they were supplied with the flour that was both their one food and their habitat.

From his weevils in jars, Swanson drew all sorts of conclusions about the growth of beetle populations. His results were remarkable in that they seemed to have none of the approximations and variability we were accustomed to believe ecologists studying live animals should expect. Instead, they were as accurate and repeatable as if they had been obtained in experiments by chemists or physicists. They seemed to say that the essential messiness of the lives of animals did not rule out the possibility that their populations could be investigated with the sort of mathematical precision that attracts so many of the best scientific minds to the study of physics.

For some of us, there was, nevertheless, always a sense of uneasiness when we thought about Swanson's experiments. They seemed to claim too much.

One day I got up the nerve to ask him, "Dr Swanson, is that really all there is to it? I mean, do you think if you keep on with your sort of experiments that one day you'll know all that's worth knowing about ecology? That is . . . will you be able to claim that the things that happen in populations of wild animals can be explained and understood from your experimental models? I mean, could you explain how a population of giraffes were – ?"

"Of course I'm not saying that sort of thing," he snapped. Dr Swanson was easily irritated. He was irritated by my questions. "What I am saying is that the experimental approach can give you more reliable data. But, yes, I do believe that in many cases you can get at the underlying biology of large, wild animals by understanding population biology in the lab. After all, all animals grow and need food and – "

At that point he lost me. I admired the beauty of his small-scale experiments, but could not bring myself to believe in their universal applicability. And I didn't like Dr Swanson very much; and I was pretty sure he didn't like me.

I wasn't alone. A lot of students were repelled by Swanson's sarcasm, by the way he dismissed their attempts to ask critical questions. But it was also acknowledged that his work – driven by his cold, keen mind, whatever the limitations of his beliefs, and his attitude – was exact and rigorous, that he was one of the few real scientists we had encountered as undergraduates.

*   *   *

But then we were getting towards the time when we would no longer be undergraduates and would have to look for jobs. Soon many of us began to realise that, even though our training had been largely inadequate to prepare us for serious biological work, there were – in this post-WW II Australia – various research opportunities opening up. So, like all the others among my fellow students, for whom biology still beckoned, I began to look for work. And for that, I, like everyone else, looked for someone to recommend me.

Percy Swanson was the only one among my teachers whose rapidly growing reputation as a researcher could be relied on to give any application of mine a powerful boost. But could I really approach him, in the expectation of mutual dislike?

I discussed this with Mervyn Andrade, one of my friends among the zoology students. Mervyn laughed. "Okay, so you don't like him, and you think he doesn't like you. It doesn't matter. Swanson's the one ray of sunshine in this joint."

"Swanson the Sunshine Boy! I don't see it. He's so arrogant and . . . unapproachable."

"In a way, maybe. But he does know his stuff, and he's a clever researcher. He's well worth cultivating."

"Yeah. But as I say, I feel sure he detests me. I don't see how I can cultivate him, as you put it."

Mervyn laughed again. "Maybe he sees you as pretty arrogant

yourself. You probably rub him up the wrong way."

"Me, arrogant? What d'you mean?"

"Well, you ask questions in a challenging sort of way. And didn't you write a funny sort of essay for him?"

"Funny?"

"Yes, about the biology of consciousness or something."

"I was sort of bored by the topics he suggested. I discussed it with him first. He told me we could write just about anything we wanted, as long as we researched it properly."

"No wonder you annoyed him."

"What do you mean?"

"Well, what he'd really have wanted would be something like what I wrote on."

"Which was?"

"Animal population asymptotes. That's on topic. Something he can relate to."

"Look," I said, feeling exasperated, "Swanson's always going on about the philosophy of biology. You know. Whitehead and Needham. The idea of consciousness is deep stuff. Why wouldn't he be really interested in it?"

"Oh, I'm sure he is. But when he tells us about these tough, speculative things, he thinks he's giving us a lot of hot new stuff. And as it's not in the textbooks, he wants us to understand how smart and widely read he is. But it's not stuff he wants us to deal with directly. I bet he thinks you're an upstart to be writing about your high flown ideas in an essay. They're the things he wants to tell *us* about. At this stage – in what he thinks is our development – he assumes they're his property, to which he's introducing us. You should understand this. You should have written about population biology."

"How the dickens do you know all this?" I said, feeling sure Mervyn Andrade was right.

"You're not very tuned in, are you, Bill? Nearly everyone gets the point about this sort of thing. Except for a guy like you who's just about as ambitious and blunt as Swanson."

"I hope, I really hope, I'm not like him. I – "

"Oh, in most ways you're not. I mean, you've got a sense of humour, and you're not stuck up and superior. But, look, don't ever work with him. That'd be murder – for one of you."

"Not the faintest possible thought of doing that," I said.

Though that was not quite true, because if Swanson's attitude had been a shade less overbearing – or if I had been more complaisant.

Chapter 2

Then we had graduated and, at a time that seemed breathtakingly soon after I had talked to Mervyn Andrade, I suddenly found I had got a full-time job as a zoologist that would enable me to get some research experience and the beginnings of professional standing. And I owed the job to Percy Swanson. I performed well enough in his ecology course, and he wrote me a letter of recommendation that surprised me with the generosity of its support.

Now I felt simultaneously relieved and tense. Relieved, because it would be henceforth possible to look forward to extended projects, tense from an ambition to get going as some kind of a scientist.

The job I landed was in Tasmania – Australia's island State, off the southern tip of Victoria, a hundred and fifty miles by sea from the mainland. I had no experience for such a position and I suppose I should have been cringing in anticipation of making a fool of myself. But I knew there were young people all over Australia at that time who were beginning scientific careers for which they were similarly unprepared.

In my case, I had been appointed to work as a research scientist on "investigations of freshwater fisheries potential." My knowledge of freshwater biology was practically nil. No one would directly supervise my work, and I would have to train myself – "learn on the job." That seems a ridiculous proposition when I look back on it, but it was an inevitable fact of life for many young Australian scientists of the time, because of the nationwide shortage of trained people.

* * *

On arrival in Hobart, the capital of Tasmania – at that time a city of about 80,000 – I was assigned a fine office and generous lab space in an old, colonial building, whose thick stone walls had once housed a hospital. A man of about thirty introduced himself as Randy Evans; he told me he would be my assistant in lab and field work. After we had talked for an hour, a dignified-looking man of middle age appeared in my office to introduce himself.

"I'm Ernest Crawford," he said, in a quiet, cultured voice, as he came towards me, hand outstretched. "You, of course, are Bill Logan. I know about you from our Head Office in Sydney. Randy here has been helping me for the last couple of years, but my need for his assistance is winding down, so he'll be able to give you a lot of his time.

"Anyway, welcome. And I hope you're going to enjoy working here. Is this your first time in Tasmania?"

Over a sandwich and coffee in his office a couple of hours later, I learned that Ernest Crawford was an Englishman who had done his Ph.D. in Oxford, and had come to live in Australia in 1938 for two years of post-graduate study in marine biology. "Trapped here by the war," as he explained, he had married an Australian, worked with army doctors on tropical diseases from 1941 to 1945, then joined FROA (The Fisheries Research Organization of Australia) to return to marine biology. He had been described to me during my job interview at FROA Head Office as the best scientist the organization possessed. Though I knew it was not his responsibility to instruct or guide me, I was comforted to know that this able and experienced biologist would be in the office next to mine.

Chapter 3

With the help of Ernest Crawford I found a small flat about a mile from the lab. I settled in with books and papers, which were almost everything, apart from clothing, that I had brought from Sydney.

On my third day, Crawford invited me into his office, where we sat for a moment in silence, almost as if to begin a prayer together. At last he lit a cigarette, drew on it with slow deliberation, and after exhaling, said through the smoke, "I should explain what I do, and where I sit in connection with what you are going to be expected to do here, Bill."

I felt myself sit a bit straighter.

"You and I are not expected to work together in any organized sense. My programme is a statistical analysis of the Tasmanian trout fisheries. Trout were brought here from the Northern Hemisphere about a century ago. That means, among other things, that I'm supposed to review all the documents that have ever been kept – those that still exist! – of all the catches of trout by sports fishermen." He smiled briefly. "I also have creel census data from present anglers to keep track of. Many anglers have complained about several kinds of deterioration in the trout populations. There are claims the fish don't grow as fast as they used to, that they aren't as big, and that there aren't as many of them as in the past."

He paused to light another cigarette.

"There are serious contradictions in the claims. My work includes reading age and past growth from the scales of trout caught by anglers over many years, to determine if there really is a slowing in growth rate. If there is, it could mean something much more serious than changes in the trout population dynamics as a result of exploitation."

"What sort of things?" I asked.

"Oh, well, some very widespread environmental or ecosystem deterioration in the rivers and lakes across the state.

"But as to my central task, determining the age and growth of fish for food or recreation is routine in Britain, Europe and North America – has been for nearly forty years. But here we know next to nothing about how fast fish grow."

He looked towards the window as if admiring the fine day outside. "Anyway, what I'm trying to do is in many ways pretty dull stuff. I justify it to myself by insisting it must be done if we're ever to know whether there's a grain of worth-while information in the records."

"Hope you won't mind my asking," I said, "but how do you come to be doing this rather than working on marine fisheries in Sydney or Melbourne?"

"Well," he said, "it's a challenge. The number of anglers has increased greatly since the war and we have to put the management of the fishery on some sort of rational basis. And the truth is that I like to work alone and this job allows me to avoid team efforts which may be very necessary but can get very complicated because of competing egos. And then all kinds of intrigues and factional disputes can happen." He gave a short cough of a smoker.

"And now I also have to explain to you in some detail – I've been instructed to do this – the project they want you to start with."

He lit another cigarette. "During the war years, when British marine fisheries were unable to operate because of the U-boats, some marine biologists started to consider the fisheries possibilities of sea lochs in Scotland. And to stimulate the production of organisms in the food chain leading to fish they added chemical fertilizers to the waters of a couple of lochs. Results were inconclusive, but seemed promising, and some people thought they wanted to try this sort of thing here."

"Sounds interesting," I said, though I really had no idea of the merits of such a scheme.

   Crawford nodded. "Anyway, some of the FROA people from Sydney came here two years ago and added chemical fertilizers to a Tasmanian highland lake. Freshwater of course; there aren't suitable marine areas hereabouts. They began follow-up studies of the lake's condition, but then they got too busy and the project was anyway deemed too expensive. But it was recently decided to get someone to spend a couple of years working on it full time, thoroughly reviewing and evaluating the results and giving us the basis of whether or not to do more of the same. That's why you're here.

   "It will be your first task here to finish this work off. I'm supposed to tell you all about it, explain the past of the project as fully as I can and transfer the data records to you."

Chapter 4

A couple of days later, I accompanied Randy Evans to have a first look at the lake that had "suffered" (as I came, eventually, to think of it) chemical enrichment. It was a small body of water, Lake Liversage, oval in shape, seven metres at its deepest, ten hectares in area. It was a lake of glacial origin, at a thousand metres altitude, in a region of dolerite mountains.

To reach the lake we drove in a FROA field vehicle a distance of about eighty kilometres from Hobart, climbing during the last few kilometres on a gravel road, past spectacular waterfalls and immense manferns, through the huge, splendidly forbidding trees of a dark, cool-temperate rain forest. We came out at last in a high narrow valley, classically U-shaped from glaciation. Here, in a wild, stark landscape of unforgiving rocks, overlain by a thin, skeletal soil, a vegetation of dark, dense and heathy shrubs clung close to the ground; and everywhere there were small snow gums with their smooth, pale, twisted limbs. The air, pure and clear, was stingingly cold. This was a zone of powerful winter influence, a zone of endurance. Around me was primeval Nature of a harsh, tough, uncompromising grandeur, looking as this part of the world would have looked before the arrival of humans.

Here, now, I told myself, as I stood in awe, shivering, I was perhaps really going to become an ecologist.

Lake Liversage was a rocky cupful of cold, crystalline water. As I first saw it, the snows had cleared from around it, but not from the heights of the steep mountain wall that loomed above the west side of the lake for another three hundred metres. The water on that day was still cold; not more than five degrees Celsius. As we rowed a dinghy out towards the lake's middle, the water seemed a light amber – a colour I was soon to associate with water draining from peaty areas. But in the case of Lake Liversage, when I dipped up a glass jar of its water, I saw that it was clear, utterly clear, with a sparkle, a natural cleanness I could hardly have imagined. The colour I thought I had been seeing was that of the mud deposited in the basin of the lake, which was visible through the transparent water.

This day was overcast, dull. But on another day, a clear, fine day without a cloud, the sky a clean, hard blue, like transparent but unreflecting glass, I climbed the steep bushwalking track that traversed the rock wall and looked down on Lake Liversage, three hundred metres below. It sparkled up at me like a great eye, blue as lapis lazuli in the sunlight.

In this National Park that contained Lake Liversage I first came to understand words like "wilderness" and "conservation", first truly felt the sanctity of nature, the interrelatedness of living things. I'm not what people call religious. I don't pretend to know what's "out there", though whatever it is, for me it's not going to be a personal deity. Since, however, we really can experience a world of wonders – still – I suppose it has to be conceded that if any being is responsible for them, it must – even if it is not a cosy presence – embody aspects of splendour. In that sense, I can profoundly revere the possibility of such a being. . . and even hope that it exists.

*   *   *

When I first saw Lake Liversage, and for three subsequent years, I was still in that state of unconflicted loyalty to science (a condition of youth and naivety) that led me to applaud anything science was capable of – except the manufacture of weapons of violence and war. By "unconflicted", I mean that when I saw what had been done to Lake Liversage I could marvel at it as an example of *scientific* manipulation of a functioning lake ecosystem, though even then it was obvious to me it had profoundly violated the lake's pristine

integrity.
     For as we crossed the lake in our dinghy, rowed by Randy, I was amazed to see great growths of aquatic plants (actually the millfoil, with the scientific name of *Myriophyllum elatinoides*). The plants were a luxuriant deep green, with whorls of tiny leaflets. They grew, evenly spaced, in four lines that crossed – two lines paralleling the lake's long axis, two at right angles to these. Each line ran from shore to shore. All these plants reached the surface, meaning that the largest were seven metres high and one and a half metres across.
     I examined them through a water telescope – a box with eyeholes and a bottom plate of clear glass – that Randy handed to me. The size of the plants amazed me. They were, as Ernest Crawford had explained, the result of paper bags filled with agricultural fertilizer – a mixture of chemicals to yield an abundance of the essential plant nutrients, nitrogen, phosphorus and potassium – that had been dropped from this very dinghy two years earlier, during four traverses of the lake, with the hope that they would penetrate the lake's soft bottom mud. Even to Crawford, who had been anticipating something interesting, this result seemed prodigious – monstrous, almost. A Jack-and-beanstalk story.
     I was dazzled by this experimental outcome, and can see when I look back that it was this dazzlement that initiated other feelings in me. Which led to my inner state of conflict . . . and I suppose, eventually, to some of my later views of science – and other things.
     For what I was trying to take in was quite shocking. By shocking, I mean the presence, the even, geometrical spacing, the abundant and flourishing vigour and size and health of these plants. It was as if, in the midst of an unsullied expanse of natural landscape, that lacked human influence, I had come suddenly upon a cultivated rose garden. The millfoil plants in this clear little mountain lake, gigantic and dominating, were – even to my city eyes – staggeringly *inappropriate*.

* * *

     I became aware that Randy Evans was watching me. "Well," he said, "what do you think?"
     "Oh, it's . . . amazing. And you must have seen it before all this happened."
     "Yes. It was different then all right. That's a fact."

After a bit, he said, "I dare say it's spectacular enough, and it's the sort of effect Dr Crawford and the others were pleased with. But it seems a hell of a shame to have done this to a nice, pure little mountain lake."

I nodded. "Well, as far as increasing the lake's productivity for fish, it may work, but . . .I can see what you mean."

"I hope you're going to be able to find out what it all adds up to," he said. "You know, there's a bunch of data and also a lot of plankton samples and other stuff still unexamined. D'you think you'll go over all that?"

"Have to," I said. "But it'll take a long time, because while I'm trying to understand what's happened, I have to teach myself the collecting and laboratory techniques I'll be using."

"I'll help you all I can," he said. He meant it, but I was soon to find he lacked the necessary training and scientific understanding. For the next six months I struggled to evaluate and understand past data collections, to gain background information by reading in the basic literature of ecology and limnology, and to plan and conduct a new series of sample collections and analyses.

Chapter 5

Ernest Crawford told me the idea of adding the fertilizer to the lake in paper bags, instead of broadcasting it over the surface, was that they would penetrate a little way into the soft mud of the lake bottom, where the brown paper would disintegrate and the fertilizer chemicals would be slowly dissolved by the lake's water mass. It had been expected that this would provide an abundant and long-lasting supply of mineral nutrients for the growth of algae and aquatic microfauna that would mean an enhanced food chain that terminated in fish.

But the bags had sunk more deeply into the mud than anticipated and much of their contents appeared to have been trapped below the surface. Enough was released to promote some plankton response, but the main result was very high, but very localised, concentrations of nutrients in the mud. Milfoil plants had always been present as small clumps scattered across the lake floor. With their even spacing and occurrence in four straight lines, the great growths now so dominant were simply responses to the locations of the bags of fertilizer that corresponded to the pattern of their original penetration sites.

The accidental nature of the prodigious growth of these aquatic plants made me think of another recent accidental, huge – and subsequently famous – outcome of a biological experiment. Rabbits had been introduced into Australia from England in the nineteenth century, and their populations had eventually spread over the southern half of the continent. Rabbits were hated by graziers because, in their untold millions, they competed seriously with sheep for pasture grasses. Soon after WW II, enclosure trials on rabbits were begun with the myxoma virus, known to be potentially fatal to them. The virus proved capable of decimating rabbit populations, but the initial continental spread began following accidental infection of wild rabbits by biting flies that carried the virus from the enclosures.

I thought too of Alexander Fleming's discovery of penicillin: another accident. And there are many other examples, as I have since learned – far more examples than non-scientists would ever suspect.

*   *   *

I began to examine the early data files on Lake Liversage and saw that a resumption of water quality analyses was needed immediately, together with plankton catches, and estimates of the abundance of the organisms on the milfoil. Later, the growth rates of trout which made their home in the lake and spawned in the rocky stream that drained from it, would have to be considered.

Now I realised I was face to face with a research problem involving a whole ecosystem.

*   *   *

Ernest Crawford and his FROA colleagues from Sydney had built an apparatus for collecting mud from among and around the root systems of the giant milfoil plants. More than a year after the enrichment they had determined that most of the fertilizer chemicals – phosphorus, nitrogen and potassium, especially phosphorus – had remained in the roots, about a foot below the mud surface. One of the first tasks now would be to see whether this was reflected in the chemical composition of the milfoil tissues.

I collected plant material from giant milfoil growths and from very small ones that, though also in the lake, were well outside the influence of the lines of enrichment. All the nutrient elements were at higher concentrations in the tissues of the giant plants. As for phosphorus, a vital element in plant growth, it was four times higher.

The water that filled Lake Liversage drained from a land-scape of dolerite – a very hard, insoluble rock. The natural levels of plant nutrients in this water were very low. In fact, there was so little total mineral content in the water that you could probably have used it in a car battery.

The result of the fertilizer had been to promote a phytoplankton "bloom" – a short-lived abundance of microscopic plants in the water, reaching levels that the lake, in the ten thousand years since the last glaciation, had probably never before experienced.

The zooplankton – of microscopic crustaceans – had also undergone a population explosion. But now the phytoplankton bloom was long past, and water analyses showed there was insufficient of the vital nutrients – principally phosphorus – still available in the lake water to promote another. Yet the zoo-plankton remained abundant compared to its amount before the enrichment. What was the source of its food that had not been available before?

I did not decisively prove it, but surmised that the perpetual dying and decay of plant tissue from the now-abundant milfoil was providing fine, particulate food for the zooplankton, in a quantity that had not previously been available.

I had a Hobart metalworker build me a stainless steel grab for seizing intact samples of the milfoil and all the organisms it harboured. A rich fauna, dominated by stonefly nymphs and aquatic gastropods (small, snail-like molluscs), was living on the plants. These animals would be prime targets as food for fish – the brown trout of Lake Liversage.

Two years after I first started work on the lake, we live-trapped some of its brown trout. We marked the fish and returned them to the lake. Later, we trapped them again and, from the proportion of marked fish recovered, estimated the total population. The gut contents of a sample of the captured fish showed that, as expected, they were eating the stoneflies and gastropods from the milfoil. I removed scales from trout that covered a wide range of sizes and ages and, from the growth rings on these, was able to derive ages and past growth rates of the fish. All the trout that had already been alive in the lake at the time when the experiment was begun showed marked increases in growth that corresponded with the beginning of increase in plankton and the growth of the milfoil with its stoneflies and gastropods.

Now I had a pretty clear idea of what the results of chemical enrichment of this small lake had been, and how it had produced its effects, even though some of the effects had been in unanticipated directions.

*   *   *

I felt something of the sense of empowerment we hear so much of nowadays. I had looked at a lake of great natural beauty, a small ecosystem but with the full complexity of such an entity and, by tracing the course of what I now regard as a sort of desecration, I had come to understand something of the underlying structure of that entity.

In fact, though I soon came to deplore the experiment in concept and practice, especially since its practical prospect as a form of aquaculture was nil, I had nevertheless been able to use it as a tool to further understanding. The ecosystem concept had now exploded from its textbook setting. I could envisage the world in terms of myriad ecosystems – as a natural *modus operandi* in ecological comprehension.

*   *   *

"But how," non-scientists might ask, "could you possibly care about 'ecosystems?'" To such a question I quickly learned to reply that scientists view the world and, by extension, the universe, according to the subject matter of their science. Their interest. No scientist's world view should be confidently regarded as likely to lead, in the near future, to an ultimate understanding of reality. Nature's face is still veiled. And likely to remain so. Anyway, as a scientist I'm not one of those with the hubris to think otherwise.

For all scientists, there are, of course, various world views that centre on their particular spheres of interest. For the ecologists, a planet that is covered with ecosystems is their view of the world. The world that is.

Chapter 6

In Tasmania you could easily get among mountain lakes, surrounded by nothing but other lakes and mountains and compact alpine vegetation . . . and feel that little had changed since the last glaciation.

You could stand on a clean sweep of beach, with nothing to remind you of the hand of man. Forests were harshly dark and alien in ways I have never felt even in the densest woods of North America, or tropical forests in Malaysia. And there were great stands of giant eucalypts that were as majestic as the venerable conifers of the Pacific northwest . . . but a lot less romantically friendly. In fact the forests of Tasmania were daunting and forbidding. And most of their trees weren't softwoods, but hardwoods – very hard hardwoods.

Not all wild places are like this. The savannahs of Africa and Australia, with their spaced trees and grazing herds of antelope or kangaroo have an heroic, golden, paradisiacal quality. In the American southwest, the monumental rocks, rich red of the desert, and blue enamel of the sky seem like the inevitable physical and spiritual home of the Indian, the eagle, the coyote. Seeing any of these places can make you feel as if you were treading the outlands of the fields of heaven.

But Tasmania was the place where landscapes were overwhelming, impenetrable, lost to the rest of the world. So many places had already been revamped, moulded according to human will. Not Tasmania. And its landscapes – whether cool, temperate rain forests, high altitude heaths and lakelands, or the clean, empty beaches and rocky coasts – were uniquely manifest.

It ought to have been beautiful, but I cannot quite think of Tasmania as beautiful. For in Tasmania, the solitudes and silences inspired a sense not so much of peace and oneness with Nature, but that you stood alone, waiting, a stranger in a place that was very old and unknowable, different from all other places. There was a feeling of presences watching from the deep scrub, that were about to come quietly out of the forest, but who thought better of it whenever you turned and might have greeted them. It was not really a sense of something threatening, rather of creatures that had been hurt too much, and now were very cautious.

Tasmania was there to be the setting for another culture . . . which had died. For genocide had been practised against native Tasmanians. The last of the fullbloods, the few survivors of earlier days when they were being hunted to extinction, died just before 1900.

Perhaps there was a curse on the place.

Chapter 7

Over coffee, Ernest Crawford and I were discussing reports I had written on Lake Liversage. They had been submitted to the main laboratory of FROA in Sydney, from which Crawford had just returned, following talks arising out of his own work of the need for revising the regulations of the Tasmanian trout fishery. "I'll tell you in advance," he said, "what you'll be hearing officially before long. The Director was impressed that you'd been able to put together a coherent evaluation of the Lake Liversage work. I think you'll find you're in considerable favour and that people are concluding that you're working like a competent scientist – in fact, like a very good scientist."

I knew I had gone red and, feeling good, started to say something, but Crawford sailed right on.

"In a way, it's worked to your advantage that you were assigned to try to salvage something from a project that had been largely abandoned. People have really noticed that you've been able to make some scientific sense out of what had been almost written off. I understand you'll soon be getting a letter from the Director that'll ask you what sort of work you'd like to do next."

"I'm not sure I want – "

"And you'll be asked if you'd like to go back to Sydney and

take part in the studies of estuarine ecosystems that are just beginning to get under way. I think the Director believes you would be an effective member of that team, among colleagues where you can develop your abilities better than here, in isolation, and I see his point. He's paying you quite a compliment, because the estuarine study's been reserved for senior people – until now."

"That's great," I said. "And it might be very interesting, but my real dream is to shake off FROA and all government scientific organizations and get into academia."

He looked at me. "You want to teach, is that it?"

"Not so much that, or even mainly that," I said. "It's more I've come to feel that I need more advanced training in ecology and that I want to chart my own course in science, and not have to follow a research plan laid down by the main lab or by Head Office."

He exhaled his inevitable cigarette smoke and nodded. "I understand. And I sympathise. But you also have to remember that biological research in Australian universities is not being given good financial support. The government labs are much better off."

"Yes," I said. "And it's wrong and unfair. Before long the universities will have to be given better support or the system will collapse. But to get into a university I'll somehow have to get a Ph.D. They really require it of you nowadays."

Crawford nodded. "I think the time's come for you to study overseas. I can write to a few of my former colleagues in Britain, and see what's possible there, if you like."

"Great!" I said. "If you wouldn't mind . . ."

* * *

But though he was a good scientist, I knew it had been many years since Ernest Crawford had been overseas, and he was likely to be out of touch with most of those he had known before he left Britain.

I wondered who else I could ask to help me get overseas. And again, I realised who it would probably have to be.

Chapter 8

It wasn't that I knew absolutely nothing of Rudolph Hauser, McIntyre Professor of Animal Sciences at Merriman University. I knew him as the author of a recent book – called "epoch-making" and "a brilliant exposition of the 'new ecology" by two of its reviewers, and other pleasant things by a bevy of others.

The book was essentially what the reviewers claimed. And it gave notice that Hauser, already well-known to ecologists from his many research papers, had now emerged as an important scientific generalizer and thinker.

But if I knew of Hauser's work, it was nearly all I knew of him – apart from a terse and brusque description from Percy Swanson. For it was he who had now got me to the United States, on his recommendation that I had just arrived in San Francisco, and was being greeted by Rudolph Hauser.

The man I met was about forty-five years old, five feet nine, heavy set but not fat, dark, balding, intense. He called for me at the bus station where I had been delivered from the airport, shook hands with a strong grip, and said, "I'm Rudy Hauser. Glad to meet you. Car's over here."

I told him it was very nice of him to pick me up and he said, "No sweat. Taxis are outrageously infrequent at the bus depot."

I nodded, omitting to add that I had hardly enough money for a taxi, anyway.

His car, to my amazement, was an open MG. There was barely enough room for my baggage. I slipped into the passenger's narrow bucket seat and we were off in a burst of acceleration. Hauser did a fast, tight U-turn and we zipped away towards some distant traffic lights.

"I feel some surprise," I said, "to be in an MG as my first ever ride in a car in the U.S.A."

Hauser laughed. "MGs aren't totally unusual in San Francisco. Sports cars are popular here. I drive one because it's small and nimble, able to get you around in city traffic. And it doesn't use as much gas, or pollute the atmosphere as much as big cars."

Then he laughed again. "I tell a lie," he said. "Basically, I got an MG for fun when I was on study leave in Britain, at Oxford, three years ago. And tooling it around on tight country roads in England was such a blast that I couldn't resist getting another one here. That's the truth! Though I usually tell people the other things to convince them I own one for serious reasons.

"Anyway, how the hell is Percy Swanson these days?"

"He's okay. Fine," I said. "And, by the way, he sends his best wishes."

As we stopped at the traffic lights, Hauser turned to me and said, "I have to tell you from day one, Logan, that I hope you can measure up to Swanson's report of you. As far as I'm concerned, you're here, to work in my group, because he said some damned nice things about you."

I felt myself reddening as Hauser went on. "I have great respect for Percy – though we by no means share the same view of ecology – and he thinks you've likely got it. And I can tell you – though you probably know this already, Percy Swanson being as blunt as he is – that he isn't given to recommending people unless he really thinks they have outstanding potential. Anyway, he's given me reason to expect a strong performance from you as a graduate student." He spoke crisply, but he also smiled.

I liked his directness. It was pointed but light. Things felt okay.

Soon he had delivered me to an apartment block where graduate students were housed, and invited me to dinner.

Chapter 9

Rudy Hauser's research group included two Americans, a Canadian, a Scot, a German and a Brazilian.   With his kind of research money he could have supported twice the number. But he was demanding of those he took on, and concerned not so much with their undergraduate grades – though most of them had done at least moderately well – but with their commitment.

He explained his attitude when we were alone in his office the next day. "I was by no means a hotshot student myself. Too many interests. And too much questioning of what we were supposed to be learning. Too much 'what is this all about and why are we doing it?' sort of stuff. Anyway, nowadays I want people who have a pretty tough-minded idea of why they are here, can tell you why in detail, and will hang on in difficult projects without a lot of ass-kicking by me. I want them to do original work, not just further versions of my stuff."

Hauser's six – now seven – graduate students were crowded into two connecting rooms. You could hardly find a place to write notes, let alone space to spread out papers or maps. The students worked in a wilderness of scientific equipment, of which there was no shortage. Hauser channelled his research money into apparatus, lab supplies and travel allowances, rather than space or comfort in the workplace. If more apparatus was needed, he had the money to buy it or have it made. Near the lab was a small, well-equipped workshop where the handier among us could make or repair some of the simpler apparatus we needed. There were three four-wheel drive field vehicles in which the group totalled about two hundred thousand miles a year.

Three of the students were away on field work, but I met Maria Schoenberg, a short, strong, blue-eyed girl, with an air of total dedication. The other students present were Pete Parkes, a native of California, and Norman Schneider, from Illinois. Norman was pleasant, intelligent, not particularly talkative, but Pete acted self-assured, and spoke freely and at length on just about any topic that arose. "Hope you like it here," he said. "Rudy's not a slavedriver by temperament. But it works out he is, because he just demands quality all the time. You have to have a really well thought out reason for doing anything when you go to him, and an equally well thought out explanation for your results. And you have to persuade him that your data are sound. That puts a lot on you, so you end up working like hell. But Rudy's a good guy. And he'll go to bat for you if you need him. Loyal to the last atom."

Too much blather from Parkes, I thought.

Maria, who was German, nodded, however, in agreement with Parkes. "That is what I think, also. Here in America is different from Europe. There, the professors are all great and you must know that. Here, the professors may be as great, maybe greater, but they wear casual clothes, and talk in an easy way. Some students call professors by their Christian names. In Europe this is un – how should I say?"

"Unthinkable?" I suggested.

Maria gave a curt nod. "Unthinkable."

As we talked, Hauser came in. "See you've been meeting people. That's good. I don't want to interrupt, but would you like to drop in at my office in about ten minutes, Bill."

* * *

"You're welcome to look around and see what the others are doing," said Hauser, as we sat in his cramped, book-lined office. "Go into the field with them if you like. I want to be sure you get started on something you'll want to hang in on and finish."

"Thanks, Dr Hauser," I said, "but if you have any ideas about what I might do, please tell me. It might be the sort of thing I'd pick, anyway."

He drummed his fingers lightly on his desk. "Well, I don't think you should try to be too accommodating to me. What you decide on is going to occupy several years of your life. But there is something you might find interesting. Something I've wanted some work done on for years. It's an aquatic ecosystem that features a thermal spring.

"There are many thermal springs at Yellowstone, in Wyoming, but there are some closer to here. What is great about hot springs is that they supply abundant hot water at constant temperature and their outflows eventually feed into, or are joined by, cold water. So you have, at the hottest, a specialised, heat-adapted flora and fauna, and nearby you have thermal gradient zones and transitional biotas as the temperature gradually decreases downstream towards 'normal'. It's one of the most interesting natural environments in any sort of ecosystem. I think a full ecological account of an ecosystem like this, with quantitative characterization of all the zones of thermal difference and their interrelationships, would really be a unique contribution. And you'd also likely come across some physiologically remarkable animals and plants as a fringe benefit."

"Sounds good to me," I said. "When do you think I could have a look at one?"

"Let's see." He consulted a desk calendar. "I'm busy with Ph.D. examinations next week. I'm afraid it'll be at least two weeks before I can get away for a couple of days."

I nodded, and we sat for a moment as he stared out his window at the brilliant San Francisco morning. "I always spend a few days with new Ph.D. candidates at the start of their work. That way I can get them launched and see pretty much what they'll be doing, so that as results come in I can make some sense of them. I also try to go out with them twice a year for a short time. I'm away from here quite a lot on projects elsewhere in the States and overseas, but while I'm here, and providing I'm not teaching, or doing some sort of administration, you're always welcome to come and see me about work.

"Between now and when we can go and see one of these thermal systems you should read up about them as much as you can."

He paused and looked at me. "You know," he said, "it wasn't just what Percy Swanson said about you that appealed to me. He also enclosed copies of your work on Lake Liversage, and, given that you were inexperienced, alone, and without a lot of modern scientific equipment, I think you did rather a remarkable job on that project. And, of course, one reason I thought a thermal spring ecosystem might be attractive to you is that it's more freshwater biology, which you've already shown ability for."

I felt myself glowing and waited for more reassurance, but he said no more on this.

But after a moment he said, "Well, you'll want to get at your reading."

Chapter 10

"Okay, tell me, Bill, what you now know about the biology of thermal springs, or of thermal biology in general." Rudy Hauser made a church roof of his hands and waited, with the hint of an inquisitor's smile.

"Well," I said, "nothing from direct experience, but from recent reading it seems that certain organisms can survive temperatures around ninety degrees Celsius for brief periods, though their maximum temperatures for long-term survival are always less. The highest temperatures to survive and reproduce in seem to be about like this: bacteria are tops at just under ninety degrees, followed by blue-green algae – which are biologically a good deal like bacteria – at about eighty; after that come protozoa and insects and maybe some spiders, and fishes at about fifty to fifty-five.

"The insects that feed on the mats of these algae are typically species of flies, and there are other flies that prey on the flies that eat algae."

Hauser interrupted. "Yes, what about those flies? How can they feed on algae at over eighty degrees?"

"They tend to pick spots where the algal mat sticks out of the water and is considerably cooler. And they have long legs that keep their bodies well above the hot algae. In places like that they can lay eggs and produce young. Oh, and there are also midges and wasp species among the insects, and there are water mites. But none of these are directly associated with the hotter patches of algae."

"And the fish?"

"None of the fish can survive more than a few degrees above forty five, except for very brief excursions into hotter places. The main thing is that there is a functioning food chain that starts with algae and bacteria and ends in predatory insects and fish. Yes, I think it could be an interesting ecosystem."

"Well," said Hauser, "you seem to have done some homework. But now please read the stuff in my book on the dynamics of production and energy fluxes in natural ecosystems. And then let's go into the field and have a look at this sort of environment. Let's say – he consulted an appointment book – next Thursday."

* * *

The system of thermal springs we visited was quite small, about two hundred yards off a dirt road and unmarked by signs. A place where tourists and other visitors would be few. It was in the Sierra Nevada foothills, past Sacramento, a couple of hours drive from San Francisco. There was a cloud of steam hanging above the springs as we approached a spot where water near boiling point was welling from the earth. Hauser told me the local name for the place was Hellhole Springs, or The Hellhole – or sometimes just Hell.

I could have camped at the springs, which were in a flattish, stony area of very little vegetation, but it was a place of the occasional cougar and rattlesnake, and Hauser suggested I spend my nights at a village called Peachtown, about seven miles away. There, for a few dollars a night I could get clean, if Spartan, accommodation, and edible food. The place was owned and run by a cheerful woman in her fifties who soon assumed an air of maternal interest in what I was up to.

"It's a strange place, isn't it, that Hell?" she said, during my first stay at her place. "Creepy, really, with all that steam. In winter, after a little snow, you'd really wonder what you were seeing. You take care out there, young man. Some of them rattlers is as big as a . . . a python. And faster than a jackrabbit."

"I'll keep an eye out," I said. "Anyway, I'll be mostly working on a flat area with very few hiding places. Don't worry, Mrs Voorster. I am a biologist, you know. I'll be watching for anything that moves."

* * *

". . . anything that moves."   I have always remembered saying that to Mrs Voorster, because only a couple of days later I first saw *her* – and I will always remember the way *she* moved.

Chapter 11

San Francisco in the 1950s: Alcatraz, The Bay, the splendid setting, the bright mild climate, the offbeat ambiance, the mix of races and lifestyles; a peculiar sweet, bright sadness always just out of reach; glory, love and sweet damnation in the Golden West.

Anyway, provincial, unhip and straight though I was, it didn't take that much prodding from Pete Parkes and Norman Schneider to persuade me it was time to sample the city at night.

Pete and Norman talked the talk of city downtowners but didn't walk the walk, and weren't in fact any less straight and hardly more hip than I. Oh, they laughed a lot and kidded me furiously. "Got to get this poor, sheltered, uptight Aussie initiated into the real, gone, lowdown Frisco."

My reply was hardly subtle, but it made us all laugh. And sometimes we would swallow several drinks after work on hot afternoons, at a run-down beer garden favoured by impecunious grad students. And then we might lurch down-townwards, gobble a hamburger and start, in Pete Parkes' favourite – and hardly original – phrase, to "case the joint".

I don't believe Pete and Norman were any more enthusiastic about the more indecorous of the available entertainments than I was. And we very soon stopped frequenting them.

And I had never been very much for poetry readings or jazz. This was apparently my great loss, considering the presence of the beatnik poets and the quality of jazz talent in San Francisco in those days.

There were also "small theatre" companies. We went to see a number of their performances, some of which were of experimental plays that seemed pretty trashy, and were performed only a few times. But at least they were cheap . . .

Anything that moves . . . Pete Parkes had dragged us to a play he said might be worth watching, adding that he knew "the gal who's the female lead." Her movements were what I noticed the evening I first saw her. Most of the cast performed badly and I cannot remember whether she "acted" well or not. Only that I was immediately and constantly aware of her movements and her physical presence. I know only that what she did was intensely vivid and dramatic, even if it was mainly just one component of acting – movement – that she used to make it so.

She moved with an effortless agility and strength. Slim, athletic, her stylish, flowing power was the kind that is seen only in those of rare physical gifts.

I lost all track of the stupid play in the first ten minutes, and spent the next two hours staring with fixed attention at this splendid creature. There is no way I can describe how her performance seemed to me – except as an epitome of energy, grace and allure.

Pete Parkes was getting a full measure of sly enjoyment out of watching me. He had already established a personal hobby of trying to infer my innermost feelings about the San Francisco scene from my visible reactions. Anyway, that's what he thought he was doing. Certainly he never let a performance of drama or music, or anything else, go by without prodding me afterwards to see what I would say or do. At the end of the play he said, with an air of as much apparent innocence as he could muster, "Could you get a lot out of that, Bill? I mean it was pretty obscure stuff. And pretty regional and colloquial. It was redolent of the Frisco scene, I guess, in a way, but I imagine to an Aussie – "

"Okay, Pete, stop the bloody clowning. I probably got no less and no more than you."

"Yeah, well, it was dull stuff, anyways." He waited, with a half-leer on his face. "But maybe the chick in the black dress had some fairly interesting moves?"

I managed a tight grin. "Yeah. Something like that."

He smirked. "Thought you might have been thinking that, old pal. Anyways, this one's from San Diego. I went to high school and college with her – Betty McMurtry. I'm going to slip down to her dressing room and sort of touch – ah – bases with her. Haven't seen her for years. Whyn't you come and meet her?"

He was grinning widely now. It was the point at which, to preserve my Oz detachment – which he and Norman weren't too keen about, and which, to keep them off balance, I tried to exercise constantly – I should have said: "Ah, no thanks, mate. That sort of caper doesn't really grab me. I think I'll give it a miss." Yes, that certainly would have disappointed the bastard.

But after two hours of Betty McMurtry, I only managed to say, "Yeah, well, why not? Sure."

* * *

She embraced Pete, though not too warmly, I noticed, while crying, "Pete Parkes, you bastard! I've been in six of these wretched little plays in the last twelve months. Is this the first you've been to?"

She half-glanced at me, who was trying to keep my interest from appearing too avid. But the flicker of her eyes told me she saw through me in a flash. And that undid me. At the same time I was filled with delight because even that almost subliminal sign betrayed – at least – interest.

It was obvious that Pete was deliberately dragging out the conversation, filling it with trivia and non sequiturs, prolonging it to tease me. Suddenly, as if he had just remembered the presence of Norman and me, he said, "Oh Christ, Betty, I'm so rude. These guys are Norman Schneider from Chicago, and Bill Logan – a tribal member of that rare South West Pacific island mob, the Australians. Bill's a real dinkum Ozzie, a fairly pleasant and almost intelligent guy, despite obvious mental, moral and physical defects."

Betty laughed. "Shut up, Pete. I never did understand how your friends put up with you." She thrust out her hand. "Very pleased to meet you, Norman. And Bill." Her speaking voice, like her laugh, was warm and musical, subtly "theatrical".

"Likewise," I said, taking her hand, which was smooth but strong, and managing to look her full in the eyes. She looked back at me and, as we held hands just a moment longer than we might have, her eyes wavered. Then, right away, she looked at me again, smiled, and said, "I hardly dare to ask any of you what you thought of the play."

Norman and Pete muttered things that were meant to be complimentary.

But I took my life in my hands, as it were ("Faint heart never won," etc.), and as she watched me, I said, "*You* were marvellous!"

"Well," she said, laughing again, a little nervously now, "thank you. Thank you very much."

"The way you moved . . . said more than a thousand words. It said . . . well, everything."

She looked me full in the eyes for a couple of seconds.

We chatted on for about ten minutes, Pete and Betty doing most of the talking as they reminisced about school and college and mutual friends. Then Pete said, "Betty, you got time to slip out for a cup of coffee?"

She smiled, then frowned. "Look, Pete, I'd love to, really, but the cast is meeting the director to discuss what we've been doing in the performances so far. In fact, I have to go right now. What a shame. Because otherwise I'd love to come."

Pete just grinned. "Sure, Betty. Some other time. I guess you'll be in other stuff after this play ends."

"Right. In fact, I'll be in a play by Christopher Fry – he's a British writer – next month. It's called 'The Lady's not for Burning'."

"That was playing in Sydney just before I left," I said. "It was a big hit."

She looked at me quickly. "Who was in it?"

"Well, there was Olivier; he was leading the group that was touring with it, and – "

"Stop right there, please," she said. "I'll quit now if I hear any more."

We all laughed.

"Are you a theatregoer, then, Bill?" she asked.

"Hardly a devoted one. But I'll go if I'm where something very good is supposed to be playing. Or if someone drags me along . . ." I stopped, embarrassed by my clumsiness.

She took it well, but I saw her disappointment and I said as hastily as I could, "Oh, look, I'm really glad Pete brought us tonight. I really enjoyed it. Particularly your part."

I could *feel* her radiant smile.

After we left, we went for the cup of coffee, and Pete said, "Not surprised she wouldn't come with us. She's a broad that was always a little bit aloof."

I looked at him. "She had a meeting. She told us."

"Oh, sure." He laughed. "That's the sort of thing she'd always say." There was an edge to his voice that made me wonder how many times he had been fended off by Betty.

*　　*　　*

Three days later I decided I was being a craven coward. But it wasn't as though I had to go on being. I had met her, and we had exchanged glances that meant, unless I entirely lacked ability to read such signs, I was not alone in at least something of what I had felt.

I didn't, of course, mention anything to Pete or Norman. I just phoned the theatre fifteen minutes before a performance and asked to speak to her.

Her voice, when it came, was quick and firm, but with a breathy, sensual quality unique to her speech.

"It's Bill Logan, here," I said.

She hesitated for a second, then, "Oh yes, Bill. Sure. How are you?"

"Fine," I said. "I was wondering if you'd like to have that coffee after the performance tonight."

"Oh, yes, all right then, Bill." But she sounded uncertain. "That would be . . . good. You and Pete – "

"This is just with me," I said.

Again the slight hesitation. Then she said, "Yes, that'd be very nice, Bill. Just come down to the dressing rooms ten minutes after the show ends – which is around eleven."

*　　*　　*

We went for coffee and a dessert at a cafe on a steep road nearby. The cafe was built out at the back so you could sit at a windowed wall that looked towards The Bay. The lighting was dim, intimate. As we waited for our coffee we found ourselves looking hard and long at each other across the small table. Her face in candlelight was angularly structured, high cheekboned, smooth-planed. The eyes were large and lustrous with a faint slant, and the mouth was firm but generous and sensual.

She met my gaze openly, not quite appraisingly, but her expression held a question. We watched each other almost warily. She smiled. "So this is kind of a getting to know you thing?"

"Yes," I said. "That's what it is. Getting to know you very well, I hope. Starting, maybe, with what makes you an actress, for one thing."

"Oh, that," she said, looking down, with the hint of a smile. "Well, there's an essay by a French writer, Albert Camus, that explains it well; I mean what it means to be an actor."

"I think I've read that," I said. "If you mean where he says being an actor can enrich your life by letting you experience other lives besides your own – or something like that."

She looked up in frank surprise. "My God! You've actually read that," she said. "I hardly know anyone, even among actors, who's read that, and here you are – a biologist – and you have."

"Well, biologists read, too. Some of them. I just happened to read it last year, the bit about being an actor."

"Well then," she said, "you know as much as I can tell you about what you get out of being an actor if you've read that. Though of course there's supposed to be other stuff, about actors' lack of confidence and their insatiable search for approval and applause."

"You confess to that?" I asked, smiling.

"Oh, sure," she said, smiling back.

"So those are the reasons?" I said. "To live the lives of others, vicariously, and to receive approval."

"Yes. That's not everything, of course."

"What is?"

"I – "

I interrupted her. "Can I tell you my impression of what you were doing, or at least part of what you were doing?"

"Sure . . ."

"Well," I said, taking a deep breath, "to watch you enlarged *my* experience, but principally my experience of *you*. Because of how you moved . . . " She laughed, blushing. I hurried on. "I felt that what I was seeing . . . was something very beautiful . . . and maybe sort of fierce."

"Oh, look, Bill, you mustn't – "

"In all truth and seriousness, I loved your performance, not because I was getting to know some character you were portraying, but because I felt I got to know something of *you* in some aspect of, well – " Here I hesitated, not wanting to sound totally ridiculous. But I had gone too far now. If I was to have any credibility I had to go on. "I felt," I said, "that you were revealing some part of your soul."

She became deeply flushed. Her poise and flash were gone. She was trying to look me in the eye, but couldn't quite make it. I took her beautiful right hand and held it, and said, "I haven't had a lot of close women friends. Not because I don't like women. But work's been paramount for me, and I suppose I haven't wanted to, well, to put it crudely, waste time."

She left her hand in mine, though I felt a small restless movement in her fingers. Then she placed her left hand over mine. After a while she spoke.

"Well, Bill, I can say quite honestly that no one's ever said things to me the way you have . . . and on almost no acquaintance. I guess I have to conclude that Australians are different." She gave a very short laugh. But then she said, "Since this seems to be a time for confessing, I felt something too, when Pete introduced us. I wasn't sure what it was . . ."

"Listen," I said, "can we get out of here? I don't know San Francisco yet. Is there some place . . . ? I know, can we get down near the water? I come from Sydney, and for me all the best things seem to happen near the water."

"We could catch a ferry and go somewhere. Maybe across The Bay to Sausalito."

*   *   *

We cruised The Bay. There was a moon backed by filmy cloud that produced a giant ring of radiance. The black water, roughened by a little breeze, slipped below us as we stood on the deck looking back at the city lights.

It was easy to embrace. Easy to hold one another as the water ran by. Easy to think, for a quarter of an hour, that this was all there was, would ever be: a salt breeze, the throb of the ferry engines, dark water, distant lights, a moon like an alien benediction. Easy to kiss.

As we came to Sausalito I thought, do you – and does she –

know what is happening here? Is this what is commonly known as sexual attraction? Is that what you feel for this stage-ranging creature of indisputable portent and power? Is there going to be more? If you drink at this spring, will it refresh and vivify you? Or will it prove to be – for you, for your type of male – damaging . . . even fatal?

The darkness was all-encompassing as we docked at Sausalito, and the beauty in my arms, though in darkness, was real. But years were to pass before what we were to become to each other would experience its ultimate test.

Chapter 12

Betty had no income except for the sporadic amounts that came from her appearances in theatrical productions. Her prospects in her chosen work were – despite her beauty and striking physical presence – far more hazardous than mine in science. Yet nothing could have convinced me that Betty would fail to make a name. Nor was I alone in this. Six weeks after we met I was waiting for her at the end of a rehearsal and chatting with Clem Mountjoy, a fat, sweating, cigar-chomping, perpetually bedraggled thirty-five-year-old – the director of the play she was appearing in. Mountjoy drove his actors with ferocious zeal. Though he hadn't then emerged as of special note, some of the actors he had directed, those who didn't actually call him a fascist and hate his guts, predicted – accurately as it turned out – that his fame was just a matter of time.

Betty had introduced me to him at an earlier rehearsal when he just stared at me, shook my hand and grunted a forced and grudging "Hi!" This time, evidently less preoccupied, he had lounged towards me as I waited and stuck out his fat hand in an arrogant gesture of greeting. He had pushed a cigar case towards me, but I waved my hand at it. "No thanks."

He shrugged. "So the boy wonder scientist is being rapidly reduced to the status of 'stage door Johnnie' – or maybe even lapdog

– by the beauteous McMurtry, eh?" He chuckled throatily, "Like many another."

I looked at him, careful not to reply abruptly, if only for Betty's sake.

"Yes," I said. "She's beauteous, as you say. And yes, I am waiting for her. And happy to do so."

I must have spoken brusquely, because Clem grinned and said, "Say, Billie, I didn't mean to get your goat. Over the last few years that gal's collected a menagerie of kids like you. It's understandable when –"

"Just as a matter of interest," I said, "would you care to define what you mean by the term 'menagerie' and the phrase 'kids like you?"

"Well, look, Billie, I – "

"Bill!" I said. "Australians, some of us anyway, tend to favour shorter, less juvenile forms of familiarity. Bill, not Billie. John, not Johnnie. Jim, not Jimmie. Dick, not Dickie. Unless you're addressing a small child."

Clem laughed roughly, loud and long, choked a bit on his cigar, blew smoke not quite towards me and said: "Oh, look, Bill, sorry pal. You Aussie males are so sensitive for such a jock-looking bunch."

"Not sensitive," I said. "Just particular."

Clem laughed again, nearly uproariously. People looked our way. "Great dialogue, Bill. You could get a job as a movie writer." He grew more serious-looking. "Look," he said, "before we start really misunderstanding one another, I have a serious question for you."

"And what's that, Clem?"

"Well, you know, guys like me don't usually get to know many scientists. And . . . I think that may be our loss. Maybe your loss too, eh? From what little I know, I guess creative work in science must have affinities at some level with what goes on in fiction or drama or painting. No?"

"I suppose that's true enough."

"Right, so why in hell do we get to know so few of you monkeys, eh? I mean would you be here, if not for Betty?"

"Maybe not. But two things keep scientists somewhat anchored as a group. Science is a sort of all-absorbing activity and many scientists don't have a lot of energy left over. And then there's this 'Two Cultures' thing that seems to keep science and the arts in separate worlds."

He was watching me now. "What do you think, personally, of the theatre?"

"I like it. What little I've seen. But the actor's point of view, well, that's something else. I mean, actors must see the world in ways that are pretty hard for scientists to appreciate. And vice versa, no doubt."

Clem puffed at his cigar and frowned in evident concentration. "Why aren't scientists more openly demonstrative about their work? Why don't they use a bit of showmanship? Surely you want your stuff to get wide acceptance."

"Wide acceptance, yes. But how wide can it be? It can be years after you publish your work before you're aware of much reaction from many other scientists. In fact, if it's highly original, it can sometimes take a generation before it's appreciated. In the theatre you expect the audience to show direct appreciation. No one to stand between you and *your* audience."

"Well, there are the critics, goddamn them!"

"Yeah, but aren't they there mainly to sort of . . . locate the work, suggest its broader meaning and so on?"

"And so?"

I took a deep breath. I guessed I wasn't getting through. "Well, the theatre public's assumed to know a bit about the history and literature that drama's got its roots in. But science has its own vocabulary and language and concepts. And not many non-scientists are familiar with them."

Clem Mountjoy held a second cigar between his teeth. He said, "I'll say this for you, Bill, for a scientist, and a young one, you have more gift of the gab than the few others of your sort I've come across. But, still, don't you think it's your job – I mean the job of scientists – to make yourselves more easily understood by the public?"

"I do," I said. "And because you know the theatre, you'll know that, in the play 'Galileo', Brecht accused scientists of having blown their chance centuries ago of making science part of the broad culture of society."

"Hey, now you cut that out, Bill! I was just gonna cite that Brecht stuff myself! How the hell do you get to know it?"

"Nothing very remarkable about it," I said. "I try to read, and it just happened that I came across Brecht's plays a year or so ago. But though I agree with him in part, it isn't just the responsibility of scientists. It's also the job of general education for people in the humanities; they shouldn't allow themselves to be cordoned off from science early in their lives and kept that way."

"But listen, Bill, give me reasons just why in hell I, for instance, might want to know a lot about science."

"I wouldn't have a clue. Or at least I can only guess you don't want to, or you'd know why. But if you did want to, maybe you'd be a better director, because you'd be working from a wider cultural base."

He laughed again, I thought a bit uneasily, but I swept on. "It's you who's asking the questions, mate, and you live in the twentieth century, with the results and effects of science around you, and you ought to understand the world view that brought them forth. Otherwise you're just like someone in the thirteenth century who sees magic in everyday events. Can only interpret those events as magic. And just accepts them as magic. I'd ask you how you can stand to live like that?"

"Okay! Of science, enough already! But don't you think the arts have got things to say that science hasn't – can never have?"

"Jesus, Clem, come on! I'm not just a reductionist. I don't believe that when you understand atoms and electrons, or reproduction in bacteria, that you understand people in their everyday acts and thoughts, or at war."

He smiled. "For a scientist, pal, you're not too much of a learned fool."

"Are scientists learned fools? Is that how you see them?"

"A lot of them. Pretty much."

"Then, Clem, the condition of learned fool is not exclusive to scientists!"

Mountjoy clapped me on the shoulder. "You're all right, pal," he laughed. Then he grew sombre, lowered his eyes and his voice.

"Let me give you a piece of friendly, if not necessarily welcome, advice, Bill."

I looked at him hard. I didn't fancy his advice. I could hardly reject it without listening to it, but suspected it would be of the sort called "meaty", and that it would concern Betty.

This proved correct.

"That gal is something. She's a great beauty. She is also a figure of power, Bill. I warn you of this. Some women are good. Some are bad. Betty is . . . what she is I can't say. But she is going to have a career. I can't be sure it'll be a huge career, though she has the force to make it so, if certain things fall in place. Betty's not ruthless, but it may seem so to someone in love with her, especially someone who isn't in the theatre and doesn't understand the sorts of demands it can make on people. Betty may let herself become possessed by her . . . purpose, her ambition, if you will. That's not so unusual for an actor. But it can make for a rough ride for others who may be close to her – maybe someone like you. So be a bit careful, feller. This gal may get to be very, very big in a few years. Ask yourself if you'll be big enough to handle it. Ask yourself how both of you would be able to handle it, if it comes to that."

"Trying to warn me off, Clem?"

"Warn you off? Not at all, pal, not at all. But trying to warn you, yes!"

Chapter 13

My major problem in getting started at Hellhole Springs was to find a way to analyse the ecosystem that would be both objective and quantitative. My work and hers made it hard for Betty and me to get together, but no matter how preoccupied I was, vivid images of her in motion were always springing up. Soon, Betty tried her best to help things along by insisting on accompanying me in my field work whenever a significant break in her stage work occurred. She had no training in science; her degrees were in arts, and from drama school. But she prevailed on me to make her understand what I was doing and why.

The first time we arrived at the springs together we stood for a while on a flat area of permeable rock – the place where the hot water welled forth. Betty regarded the steam clouds without too much interest. Hot springs were quite common in California and she had seen a number of them in her time.

"Anyway, Bill, exactly what is the cause of a hot spring?" she asked, as she helped me unload equipment from the field vehicle. " I mean, just why is the water hot in the first place?"

"Hot springs are found in a number of countries," I said. The more spectacular ones are those that periodically erupt; they call them geysers. There are more than three hundred geysers in

Yellowstone, which is over half the world's total. This one we're going to work on doesn't erupt, is therefore not a geyser – and is not spectacular.

"But as to what *causes* a hot spring, it's the heating of groundwater that percolates through permeable rocks, or through fissures in impermeable rocks that have split apart. As this groundwater comes in contact with the terrific heat of molten rocks, or magma, of the earth's interior, it gets hot."

"How hot is this molten stuff?" Betty asked.

"Bloody hot! A thousand feet down in a thermal region the rocks are already heating the water to boiling point.

"Some hot springs have colours on the surrounding surface rocks from the hot water that splashes on them. The colour mainly depends on the particular chemicals dissolved from the rocks the water passes through. And sometimes the water of a hot spring itself looks very highly coloured near where it emerges from the earth, but that may be due to the abundance of heat-tolerant blue-green algae that often occur there."

I thought (not without a slight self-satisfaction) that Betty's facial expression suggested not only interest but even, perhaps, a touch of admiration at my knowledge. I did not bother to mention that I had known almost nothing about hot springs until my very recent study of the relevant literature at the urging of Rudy Hauser.

* * *

To start my study of the ecological processes in the ecosystem of Hellhole Springs I had decided to utilize the techniques of "production dynamics" and its related field of "community energetics" – approaches that were just then coming into use by ecologists.

These involved somewhat difficult concepts to get across to a non-scientist. But Betty was resolved that I must try, and that she would do her best to absorb what she was already calling "jargon."

I attempted to shoulder my task.

"Let me try to make a few things clear," I said. "Everything we do, everything we are, everything that lives in this world, is driven – in all the processes of life – by the energy of solar radiation. These emanations from the sun not only supply the heat that gives our atmosphere and oceans their temperatures, but also supply the energy for plants, through the process called photosynthesis, to make proteins, carbohydrates and fats from the molecules of nitrogen, carbon, hydrogen and oxygen of the atmosphere and the oceans. And then, animals that eat the plants, or that eat other animals, can, by digestion and absorption and assimilation – which are pretty complicated biochemical processes – eventually incorporate the proteins, fats and carbohydrates into their own bodies. And from these substances they build their own organs and tissues during their growth, and maintain those tissues throughout life against breakdown and damage."

Betty looked at me a bit uncertainly. "I think I know a bit of that . . . sort of," she said.

"Okay. But, you see, part of the solar energy that lets all this happen gets returned to the non-living world all the time a plant or an animal is alive, because every process in an organism demands energy. But an amount of energy equal to what is used is eventually released, but as heat, not light. And organisms break down carbohydrates or fats – or even proteins, though in the case of animals they're the main building blocks of our bodies' tissues, including blood and muscle. They break these substances down to get the muscular energy to search or hunt for food, or just to maintain body heat. And the energy, once used by these processes and activities, gets released to the universe again in the form of heat. And when we die, microbes and a whole battery of little critters go to work reducing our bodies, first to dust, and then . . . in the end . . . to nothing much at all. Except that this final breakdown of the tissues once again results in heat liberation."

"And that's it, then? The end of it, I mean?"

"Yeah, well maybe from the viewpoint of the way we live our everyday lives."

"There's more?"

"Well, as our tissues decompose after death, they liberate the elements – the carbon, nitrogen and hydrogen and oxygen back into the world."

"So it's a cycle?"

"Yes, in a sense. But the energy that started as solar radiation is eventually all released as heat – and that can't be used again."

"Interesting, I guess," she said. "But isn't this sort of medical stuff? I mean, what's it got to do with ecology?"

"Well," I said, "just to round out the picture, once the elements are released into the world again, they can never make their way back into the bodies of living creatures to become parts of tissues or organs again, except by again entering what we refer to as a food chain that starts with plant photosynthesis. But always only a proportion of the organic substances produced by plants can end up as a somewhat 'permanent' part of the organs in the bodies of animals."

"What do you mean by a proportion?"

"I mean that if a growing animal eats . . . Wait a second. Look over there, way off to your right! See?"

"See what?" she said, turning her head too far.

"No," I said, pointing. "Up there on that ridge about four hundred yards away."

Betty stared hard, then turned back to me, her mouth half open. "Is it . . . a cougar?"

"It certainly is." I moved to our field vehicle, about twenty yards distant. I pulled a pair of binoculars and a rifle out of the rear seat.

"I'm not going to do anything about it unless it comes towards us and shows real interest. Which I think is most unlikely. It's a potential danger, and now we know there's at least one hereabouts we'll need to watch out, but, you know . . ." and I laughed.

Betty looked at me sharply. "What?"

"Well," I said, "we're looking at the end of a food chain." I pointed again at the cougar, which was by now sharply etched against the blue sky. "That creature has to eat many times its own weight in food to get to its adult size, and because all the big cats are the ultimate predatory forms– I mean they eat other animals almost to the exclusion of plant material – you're seeing, as I say, the very end of a food chain, alive and kicking."

Betty stared hard at the cougar, screwing her eyes up against the bright light. I thought she looked fairly impressed.

"There's another thing that's kind of interesting," I said. " If you bring me just the skull and jaws of almost any mammal, I can probably tell you where it fits in a food chain, even if I don't even know exactly what species the animal itself is, or where it came from in the world."

She looked at me – a bit wonderingly I thought.

"I mean," I said, "I'll tell you whether it's a plant eater, or a predator eating almost nothing but other animals, or an omnivore – a mixed feeder taking both plants and animals."

"Oh, Bill, surely you can't be that omniscient," she protested. "Hasn't a predator got to be fast and active and . . . and *want* to catch other animals? Won't those things be more important than anything else?"

"Sometimes . . . But basically, if you don't have the specialised teeth and jaws to be a successful predator, you won't be. You can't be. A specialist predator – and a cougar's a great example – has dagger-like canine teeth to kill its prey, sharp blade-like incisors to bite off pieces of flesh, and pre-molars in both upper and lower jaws that move against each other in a scissor-like way to cut tougher tissues like tendons and even bones. But it won't have molars to grind food. It'll just swallow it in lumps, gulp it down. If you've ever had a cat or a dog, try to remember how they ate. And predators will have no tall, hard-wearing molars such as you'll find in deer – or in sheep or cattle – that can grind dry vegetable food. Their types of teeth – or you can refer to them as their 'dental formulas' – determine animals' positions in the food chain."

"What about us?" she asked. "We hunt and kill and eat other animals, but we don't have these dagger-like teeth you're talking about."

"Good point, love," I said. But see, we're a very particular exception, because we can kill much larger and more powerful animals, even gigantic animals, because we long ago developed special hunting strategies, and we also make and use weapons. But we could never have killed large animals with our teeth. Basically, we're still the omnivores we evolved to be, in the middle of the food chain, able to kill and eat very small animals because we do have small canines and incisors to cut plants and flesh, and are also able to eat fruits, roots, grains and nuts and other plant material with molars that, though small, are capable of grinding the softer kinds of vegetable food."

"So our real ecological place is as potential food for things like cougars," she said.

"Sure," I said, "except that our smartness and our weapons have combined to make this unlikely. It's we that are the threat to cougars. I mean unless we get either very stupid or very unlucky."

We watched as the cougar moved around slowly, then slipped out of sight over the ridge.

I saw Betty shiver for a moment.

"Anyway," I said, after a pause, "most of the matter and

energy an animal eats is 'wasted'. For every pound of food eaten, only a minor percentage can be retained for growth or to replace tissue lost from wear and tear. Anything from a little to a lot of heat energy is released as a result of the 'burning' that's metabolism – which involves all the body's processes, including activity."

Betty gestured at the thermal springs, out of which clouds of steam were issuing. "But I still don't see what all this sort of stuff's got to do with your looking at the bugs that live in this."

"Well, all the creatures in this spring form a unique assemblage, part of a special, heat-adapted ecosystem. And every ecosystem has its own features. If you want to, you can describe those features simply in terms of the physical characteristics of their organisms. The creatures on a sea shore are very different from those in a desert. An Arctic ecosystem, with mammals as its top predators, resembles neither a seashore nor a desert. Okay?"

"Of course, but – "

"But," I said, "the little stream running from this hot spring has many creatures that are much the same in form and bodily structure as those coming from an ordinary stream. I mean, there are blue-green algae, and insects and fish that could occur in many sorts of streams other than this one."

"And so?"

"And we can establish differences between these organisms by concentrating on the rate at which they use solar energy, and the nutritional substances they make and utilise in the course of their lives. Plants and animals living at higher temperatures can be expected to be doing all this much faster."

"How much faster?"

"Oh, the rate may double for every ten degrees rise in temperature."

"So," she said, "if you were looking at these same sorts of bugs in a cold stream and then at – what's the temperature here?"

"About ninety Celsius at the spring, dropping slowly as we go downstream."

"So the rate at which these things happen could be enormously faster here where we're standing?"

"Well, that's what I'm expecting I'll find."

"But how can you measure that?"

"Just help me collect some of these things you keep calling bugs and I'll show you," I said.

We gathered selected organisms from several points along the course of the stream that flowed from the thermal spring and then placed them in small containers. Betty seemed quite interested in this, and affected none of the disgust many non-naturalists display when handling 'bugs'. From these containers I bled off small water samples for determination of oxygen content. I then asked Betty to put the sealed containers back in the stream at the places and the temperatures from whence they had come.

"What's this for?" she asked.

"Before we leave here in two days," I said, "we'll take another water sample from these containers and measure the oxygen content again."

Betty looked at me. "Won't it be lower . . . a bit?"

"Exactly," I said. "And that means we'll be able to calculate how high the respiration rate of these creatures is at various temperatures along the course of the stream. And we'll also sample the performance of similar organisms in a small cool stream that joins this system a few hundred yards downstream."

"But how will this relate to what you've been telling me? I don't . . . "

"The thing is," I said, "with all those functions and processes we've been talking about, the energy needed to produce them will be supplied to, and used by, organisms at different rates depending on environmental conditions, and –"

"Yes, I get that, you've already explained it."

I could feel that Betty was getting a bit impatient and, as I handed her more glass containers to put back in the stream, I tried to hurry ahead. "Well, most energy transactions in the bodily processes mean – in biological terms – something is 'burned', usually requires oxygen. And as you know, all burning produces an amount of heat – heat energy. Okay?"

She nodded.

"So we can say that organisms – while they are alive – are involved in an 'energy flux', and that the amount of energy liberated as heat in biological 'burning' bears a very close relationship both to the amount of oxygen they must use and the amount of body substance that is consumed."

Her eyes lit up. "I get it! So you can measure oxygen used over some time period, and you can somehow also use this as a measure of this 'energy flux', and of body substance used up."

I felt good at last; she was getting it.

Suddenly she flung her arms around me and kissed me. "I only *just* get it, Bill," she cried. "But I do get a sense of your enthusiasm for it. And I can see that it really must be interesting to look at a world of all these creatures in a way that most of us would never ever imagine."

"Yes, it is," I said. "But if it's often fun it can also be very strung out and tedious. What we are doing here and now is just a thousandth of the sort of thing I'll need to do to get a reasonable picture of this kind of ecosystem – which has very extreme environmental conditions. But when I have that I will know things about the essential biological nature of it that I couldn't possibly have got by other means."

I thought she looked doubtful, so I said, "Look, even a hundred years ago, thousands of competent naturalists could have identified and given you an excellent scientific description of the 'bugs' in this stream. But as they aren't all that different in type and appearance from the 'bugs' in many streams that have completely 'normal' conditions, what else could they say, except the glaringly obvious: that somehow, the ones in hot streams have the ability to tolerate very hot conditions? Nowadays we can take this a whole stage further, by beginning to understand just what the environment demands of them and the nature of their physiological responses to this."

"Okay. Great!" said Betty. "But also – and still – so what? Why do you want to put so much time . . . and ingenuity . . . into studying this kind of thing?"

I looked at her, wondering how to respond. At last, I said, "I guess it's a bit like medical science. Here's the body – this ecosystem. I know it's wonderful and enormously complicated. But that's why I have to analyse its functions and processes and try to understand them. We can't just let things sit like that. The whole world, the universe, is out there, and if you're a certain sort of person, you want to understand. So here's this body that's a particular sort of mix of plants and animals and microorganisms, all interacting and functioning in various ways, and some of us can't just leave it at that. We have to try to find out how things work."

"Yes," she said, a bit doubtfully. "But will you ever be able to say *why* they work?"

"You're getting into metaphysics now," I said. "I'm trying to become a scientist. Mostly we don't like to ask the 'why' questions. And if we do, we're usually getting towards the end of our scientific lives." Then I stopped, cancelling the grin I felt sneaking across my face.

"You're right, of course," I said. "I'm always asking myself the 'why' questions. Sometimes answers at the 'how' level lead on to answers at the 'why' level. But not often. But I think as time goes on many scientists are hoping that they will eventually be able to confront things like purpose and ultimate cause. Meanwhile we stay with this drudgery – just keep on digging."

"How long will this field work here take you?" Betty asked, while glancing quickly up towards the ridge where the cougar had been.

"Well, certainly until any significant snowfall occurs that would prevent me getting in here. And if the snow's light, I'll work right through."

"You mean, for a whole year, through all the seasons?" She began to look anxious.

Seeing this, and not wishing to answer directly, I told her that the next step was to determine how much algal food the flies ate, and how many of the algal-eating flies were, in turn, consumed as prey by other flies. Later would come estimates of the rate at which fish – which didn't stay in the hotter locations but entered them briefly to feed – consumed the flies and other insects and invertebrate life. And in the end I would also be tying all these events and processes together with the kind of techniques we had been working on together.

She was incredulous. "You mean you could be at this field work for a year or more?"

"That's about the size of it," I said. "There has to be a solid quantitative base if I'm going to come to any conclusions later."

"But, good God!" Betty cried, her slanted eyes flashing, "how can you bear to go on doing the same thing like that?" She turned away from me then, and in one regal movement strode twenty yards along the edge of the stream, discontent in every quick, impatient swing of her arms. Then she wheeled, suddenly but with that remarkable, inherent grace that shone in every movement she made, and came all-but-running back to me.

I almost thought she might run at me with violence in mind. I put up an instinctively restraining hand. "I have to," I said. "And I want to, because that's how I'll get any decent science out of it all.

You know, I don't ask you why you have to do umpteen rehearsals to get a part right. I know you have to."

"But for Heaven's sake, Bill, you're talking twelve more months of repeats, not a couple of weeks."

"Yes, well, that's how science has to happen. There's no short cut."

"It's dull then," she said, "deadly dull. That's what I feel about it."

"I understand. And what you do is grand, and vivid, and changeable. But don't forget, Betty, if you get into a hit play, or films, there could be dozens, scores of repetitions of the same material. What about that?"

She declined to answer, dismissing my remarks as a red herring, with an imperious, jabbing hand gesture. Then silence.

* * *

Well, I had tried. Perhaps she would take in just a bit of it all. I reminded myself that she had absolutely no education in science. But on the other hand, she had showed interest and understanding, until I had spooked her by indicating how long things were going to take. Maybe some other time, some time when she had got used to what I was doing for a living, she would become more tolerant of the way I needed to work. I would just have to keep hoping.

Meanwhile, I knew that, if I was in earnest about getting on in science, nothing must divert me. And Betty would have to become reconciled to my absences. This was not easy for one who, by her appearance and the powerful allure of her movements and personality, had already become accustomed to dominating her friends and fellow actors. . . and a couple of past lovers. In me, she now saw one who was drawn to her but would keep the central themes of his existence intact.

It was in fact my work that held me from becoming absolutely besotted with her.

Chapter 14

In my second year at Merriman I began to sense what I wanted in an eventual career, and it did not include tarrying over the Ph.D. for more than the minimal time required. I worked at a furious pace.

The work plan I developed with Rudy Hauser's approval was to obtain a quantified accounting of the ecosystem of Hellhole Springs, to include ecological conditions in the hottest parts of the springs and also in the mainstream further down where the temperature had declined to normal. I also sampled from locations in the gradient zone in which types and numbers of organisms were transitional between those highly tolerant of heat and those of cooler waters.

Fifteen months after arriving in San Francisco I already had a year-long record of the production dynamics of these waters. I had measured rates of production of organic matter by algae and bacteria and their accompanying energy flows near where the thermal spring issued from the ground at more than eighty degrees Celsius, and had done the same for invertebrates and fish. I had also compared these processes in the main stream well below the thermal spring's influence, and in cool tributary streams whose organisms lacked all signs of special thermal tolerance.

Progress had been satisfactory, but if it seemed unlikely that the results, when published, would be considered a fundamentally important advance, I believed they would be judged "interesting", if only because they comprised what was, at that time, an almost unique account of the production dynamics of a thermal spring considered as an ecosystem.

It was not that my results would never be improved on. They surely would be, though not by me. I believed the study would serve my purposes if I laboured on it for another year. By then I would have secured a workmanlike body of information and be able to present a sound preliminary study for my Ph.D.

From the jump, my interest in Hellhole Springs had been twofold. First there was an opportunity to extend my experience of the structure and functional web of an ecosystem – experience that began with Lake Liversage. And this "period in Hell" was not a mere repetition of the Liversage work but was smaller in scale, more detailed, quantified, employing state-of-the-art techniques. It also had the benefit of being supervised by Rudy Hauser. Second, I knew the advantage of studying an ecosystem with features as exotic as those of Hellhole Springs. These features alone would automatically attract some attention to the work on its publication; I would be assured of an audience for my work.

I was sure my experience in developing concepts and techniques for studying aquatic systems would serve me well for investigating ecosystems of all types. They were the sort of background needed by every versatile ecologist. In my mind's eye was a slowly forming image of the course a future career could take. The image stimulated me . . . And it troubled me, because I knew of no institution or agency that would give a person as young as I the scope to do what I was beginning to dream of doing.

*   *   *

Pete Parkes and Maria Schroeder were the dominant personalities among Hauser's graduate students. Pete was working on a single species – the bald eagle, a somewhat unusual topic for Hauser to be supervising. His project, already in its third year, was to study the eagle's population biology – its reproduction, movements, position as a top predator in its ecosystem, or rather in the several ecosystems of which it was a member. The huge mobility of these legendary birds ensured that any individual of its species could readily include many ecosystems in its feeding range.

Pete – even if half-jocularly – tried to make people feel he enjoyed a special relationship with Hauser, that the glamorous object of his studies gave it automatic status: "There's only about three ecologists anywhere in America seriously studying a U.S. national symbol," was his smug assertion. To be working on this noble animal was, Pete Parkes claimed, evidence that Rudy Hauser favoured and looked out for him. "Go figure, Logan," he spluttered. "I study eagles, man, eagles! You study the bugs in a hot water pipe. Christ!" And he laughed like a loon.

He rarely missed a chance to hint at relations between Betty and me. "How's Lover Boy Logan?" he would chortle.

"Y'know, pal," he jeered one day, "you owe your entire emotional life to my kindness. I mean *kindness!* How else is an Australian – an Antipodean hick – gonna get a gal like Betty McMurtry when a hunter of eagles is around, unless the eagle hunter's in the middle of "Kindness to Third World Natives Week?"

"Does it hurt a lot, baby?" I said, and threw him a laboratory wash bottle. "Here, diddums, lie over in the corner and suck on this till it's time to change your nappy."

Another time I said, "You know, mate, if brains were ink yours wouldn't make a full stop." The trouble was it wasn't always just banter. I realised I somewhat disliked Pete Parkes and that he had it in for me. It dawned on me that he must be jealous, really had lusted after Betty when they were in college, and that now he was faced with the fact that she was actually living with me.

One day he irritated me more than usual. "You look relaxed, Billy-Boy, as if you've had a very nice time. Maybe a nice bed-time?"

"My name's Bill," I said. "I don't permit liberties with my name by ignorant twits."

"Sooooo sorrreee, William, lover. Anyway, how's the hotsy-totsy Betty this fine morn?"

"Why, you sex-fixated Yank idiot, don't you just shove it?" I enquired.

"Don't get your shorts in a knot, sweetheart," he grinned.

"Well, I guess they must be already."

I moved quickly, then, in his general direction. He rose from his seat. Other graduate students turned towards us.

"Something I'd like to tell you, Parkes," I said, as coolly as I was able. "If you can spare a minute from your profound ponderings. But let's step outside where there's more room, less equipment to be blundering into."

"Sure," he said, pushing his seat back. "Any time."

I had the impression he had been wanting this. And then I had a second impression that, now it was happening, he was not so sure.

Outside the building, in a fresh morning breeze, we faced each other.

"I'm not a psychiatrist," I said, "so I can't tell why your head is sick."

He grinned in a wolfish way. "Physician heal thyself!"

"Yes," I said. "No doubt. In the meantime, I think I know the cure for *you*."

"Which is?" he said. He stood up straight. At six feet four he was four inches taller than I. And now, at last, his face was still, watchful.

"Ah," I said, "now you're asking. But I'm not sure you want to know. Let's put it like this: the remedy's a secret. But I could let you have it at any moment. When I decide to administer it I'll tell you what it was. Afterwards. But don't fret. You'll never even know I've given it to you. For now, you'll be able to help yourself best by just making sure you stop vomiting any more of the muck you've been sicking up. I mean, Pete, old mate, just shut your bloody trap!"

He looked at me for a long moment. He squared his shoulders. I thought he might move then, but he didn't. Instead, a look that might have been regret washed over his features. "You're taking things kind of seriously, Bill, aren't you?" he said.

"Very seriously. Glad you understand."

"Well," he said, "I think we should forget all this."

"I hope you do," I said, as I turned my back on him, and went back to the lab.

He followed. The silence in the lab was so complete that I realised things must have been in a hubbub immediately before our return.

It was the end of his nonsense. We never became close friends. But civilities prevailed from then on.

*   *   *

I found Maria Schroeder good company once her initial stiffness of manner – due to shyness – wore off. She was an early feminist, fiercely dedicated, not so much to science as to success and recognition in what she regarded as an intellectually tough and demanding profession. Her research was on a marsh ecosystem, with emphasis on competition for food and space between several bird species. Soon, I began to feel sorry for her and after a few months I dared to tell her why.

It was one evening. I had been working in the lab till ten or more at least three nights each week. Maria worked longer than any of us. She was always there till near midnight, sometimes later. She spent a lot of time running and re-running slow motion film she had shot of the supposed competitive behaviour between bird species. Once she was sure she could accurately identify the various stages of this behaviour, she used the observed criteria to make counts of such interchanges in the field.

She looked tired and wan that evening and I said, "You're doing too much, Maria. You'll kill yourself. Take a break."

"There's a lot of work, Bill," she said solemnly. "But I have to finish. I'm thirty. I shall be thirty-five before I will get a Ph.D."

"What'll happen then?"

"I will try to get a job. It can be not so easy for a woman. But with this qualification I may be able to. A solid academic position for a woman is – "

"Can I say something?"

"Of course."

"To be blunt, I think you've probably got more brains than most of us, maybe than any of us."

"It is nice of –"

"And yet, your research project is pretty mundane stuff. There are scores of people in North America – perhaps hundreds – that are seriously studying marsh bird populations. With your ability you should get Rudy Hauser to give you something more challenging – something more unique. You're smart, Maria. You could go a long way in ecology."

"I chose the work myself, Bill. Professor Hauser wanted me to study Hellhole Springs. Or the biology of mountain sheep in their natural ecosystems."

"And so you should have, by God!"

"I didn't want to fail. I have not your self-confidence. I wanted something I could do very completely. Something that enough other people have an interest in, so that many would be able to appraise my work and recognize it as good of its type. I would like to secure a good academic position."

"Listen," I said, "you don't want a lot of people who'll think your work is 'good of its type'. You only need a few top class ecologists who know it's really good stuff. There isn't a chance that anyone with your ability and powers of work would fail to make a good job of a tough and original project. You're selling yourself short – in my opinion. You shouldn't be worrying about a meal ticket, a 'position'. You'll get all that. But now you have one of the best ecologists around as your supervisor. He'll entertain your dreams as long as they're not ridiculous. You should be shooting for the stars. With your ability you'd get there."

She laughed. Then she frowned. Her serious blue eyes stared at me from under straight brows. In her sober nordic way she was a good looking girl.

"Perhaps you are right, Bill," she said. "But you are not a woman. We cannot afford to take so many chances."

Chapter 15

Betty and I had found it true that we could live more cheaply together than singly. Our less than modest third floor apartment – a bedroom, a sitting room, tiny kitchen, tiny bathroom – cost us only thirty percent more than I had been paying for a bed-sitting room with stove and sink, plus shared bathroom. But though we had settled into what approximated to our version of domestic bliss, Clem Mountjoy's prophecy suddenly began to be fulfilled.

A member of the theatre audience that had attended Brecht's "The Caucasian Chalk Circle", in which Betty was the female lead, came to see her after her third performance in that play. A tall powerful-looking man with dark, hawklike features and dashingly cut casual clothes, he shook hands with her and said, "Irve Robichaud. Found your interpretation fascinating. I represent Cosmo Film Studios in Hollywood. Wonder if you'd have any interest in possible work in movies?"

When she told me later that evening I felt a blow to the pit of my stomach. I looked at her, at the eagerness, the liveliness, the excitement dancing in her eyes. It wasn't possessiveness I felt . . . Well, yes, I suppose it was. But it was the possessiveness that says, "I had hoped for a little time while we were young when we could be entirely together, unbothered by other matters apart from unavoidable

work. A time when we could luxuriate in being completely
preoccupied with one another." Now I knew she was getting into
something that might cut hugely into our lives together. Of course, I
was conveniently ignoring the fact that my own future work could
also be extremely demanding . . .

I managed to rise to the occasion. "You're on your way,
Betty. You know that. I know it. Nothing will be able to stop you."
I tried to keep my tone positive.

"Oh, Bill," she said, seeing through me, "it'll be okay."

"Of course, my dear, I know that. Just what did this Irve
person say? I mean, exactly."

"Oh, not much. He said he was excited by . . . by the way I
moved . . ."

"Said that did he?" I managed a laugh. "The lecherous
bastard. Anyway, tell me something unexpected . . . or less
predictable."

She laughed too. "Well, he said my voice would work well
in films, and – "

"And?"

"And that my face had the sort of angles that film better than
a rounder, softer face, and better than – "

"Better than a face that isn't as beautiful," I said. And kissed
her.

*   *   *

Even if she did get into films we assumed it would be
unrealistic to anticipate quick success. The studios sized up very
carefully those people they finally decided they wanted, and that
was supposed to take some time. Of course, there were famous
exceptions.

Screen tests were the first serious thing. A week after Irve
Robichaud introduced himself she received an invitation to come to
Hollywood for tests.

I told Rudy Hauser what was happening, and asked if it
would be okay with him if I took a few days to accompany her to
Hollywood. He had met Betty. He was enthusiastic. "My God! Bill,
that's exciting stuff! Sure, go with her, of course. She'll need support.
Wow! That's lots of fun. Have a great time, both of you."

We drove down in Maria Schroeder's old Dodge, the loan

of which she had insisted on. She had met Betty, was timid in her presence, awed by her beauty and personality. But now she said, "It's wonderful for Betty, Bill. You must take care of her. It will be a big strain as well as a big excitement."

* * *

At Cosmo, Irve Robichaud embraced Betty. In a manner and tone clearly intended to be welcoming, masterful, overwhelming, he said, "Wonderful to see you, my dear. Welcome to Hollywood. Enormously glad you decided to come. Your test will be the day after tomorrow. First, Rolph Hammersmith, Cosmo's Vice-President, wants to meet you at a small lunch, with a few senior executives. It'll be in his office tomorrow."

He turned to me, plausible, urbane. A large, smooth hand shook mine. "You must be Betty's friend. She told me about you, said your name was Bob, right?"

"Bill," I said. "Logan."

"Of course, Bill. My memory. Terrible. Getting worse every year. Every month. Good to meet you. Thank you so much for delivering Betty to us safe and sound."

He glanced back at Betty. "Know what? While you're meeting with Rolph Hammersmith and his team, Bob – ah, Bill and I will get lunch at the Studio Restaurant. That's where everyone eats. Lots of actors usually there. Would you enjoy that, Bill? We'll likely see people like Luke Westlake and Constance Franklin, and, oh, Josh Fuseli . . . lots of others. Appeal to you?"

"Of course," I said. "Thanks very much."

"Absolutely no trouble. A pleasure, my dear boy." He was sounding slightly English, now, but I guessed he was probably from somewhere like Poughkeepsie.

* * *

We slept poorly. Betty had earlier mislaid the habitual poise which, to my great admiration, she had thus far retained. Late that afternoon, in the hotel room the Studio had reserved for her (I had been thoughtfully booked in on another floor), she suddenly started to panic. "I can't do this," she gasped. "I can't do film acting. I'm not any sort of real actor yet. I'm years away from being. All this is too early, Bill. I want it, but I haven't earned it. If I mess it up no one will ever give me another chance. I – "

"Cut it out, Betty," I said. "Look, as far as I've ever heard anything at all about this kind of thing, these bloody talent scout types – and that's all this Irve character is, for all his slick performance – are always bringing in likely looking kids. A few succeed, a lot fail. Most of them aren't actors of any description. More likely to have been assistant golf pros or perfume counter salesgirls. As for you, kid, you've got all the looks and style and personality that are allowed in a female of your age. You have acting experience. And you're intelligent. If you don't make it right now it won't be because you aren't better prepared than a lot that do make it. It'll just be . . . luck."

* * *

After Robichaud had ceremoniously escorted Betty to Rolph Hammersmith's office for lunch, he took me to the restaurant. We located an empty table and he ordered a martini. "What's your pleasure, ah, Bill?" he asked, smiling widely. I asked for a beer and was brought something that looked like beer but which I decided was too weak and watery to be taken seriously. Oh well, it was cold and wet, and I was thirsty.

Irve Robichaud sighed as though exhausted and offered me a brown cigarette from an engraved gold case. I refused. He shrugged, took one himself and lit it with a matching gold lighter, shooting snowy white cuffs with gold cufflinks as he did so. It appeared that the cufflinks matched the cigarette case and lighter. I found myself looking at his gold tie clip . . . and sure enough. I looked him over as I raised my glass of "beer". His dark suit was impeccable, the jacket wide at the shoulders and just slightly pinched at the waist. Hollywood-Italian, I supposed. A spotless handkerchief flowed from his breast pocket.

He ran a careless hand through rather long, but perfectly barbered, grey-flecked hair. He glanced at me then, the glance sharp, questioning. I thought it did not imply interest in me, but that he was just trying to "place" me. He blew smoke with an air of careless elegance.

"We should order some lunch, er, Bill," he murmured. And signalled a waiter. After we had ordered, he looked across at me with a more genuinely friendly air. "You know, Bill," he said, "that young woman will go a very, very long way."

"Glad you think so," I said.

"Yes, she has it. I'm not often wrong and I've never been more certain about anyone." He narrowed his eyes a bit, watched me through smoke from his brown cigarette. "If I may make so bold, what are your . . . expectations?"

"Our expectations?"

"Well . . . exactly."

"I think we think we'll marry. Not sure when."

He gave a just audible sigh, tapped his finger on the table top and flicked ash accurately into a large brass ashtray.

"I won't conceal it from you, Bill," he said. "I'm relieved you're only *contemplating* marriage."

That wasn't quite what I'd said or meant, but I let him continue.

"Because, with what Rolph Hammersmith may lay before her – I'm assuming her screen test will be okay – both of you might want to think very carefully about the implications for your futures.

"I mean your immediate futures," he added quickly. "Naturally you'd be free to do whatever you thought fit when things have had time to shake down."

"Well, that's good to hear," I said. There was enough irony in my tone for him to get it.

He stared at me. "Look, Bill, I'm speaking here as a friend. When people begin a film career – as may be about to happen here – things can get tricky for all concerned. In particular, the area's a graveyard of disappointments and broken promises for those who are, as it were – "

"Bit players?" I said.

"Bit players? What d'you – ?"

"I mean hangers on," I said. "People like me."

"Please, Bill, let's not be crass. You seem like a decent young man. My only idea is to warn you. I've seen lots of needless heartbreak in twenty years in the movie business."

He actually looked as if it mattered to him. But to me his remarks seemed impertinent, considering he had only just met either of us. It was interesting, though, that this was not the first warning I'd received.

"Thanks," I said. And I repeated what I'd said to Clem

Mountjoy, "Are you sure you don't mean to warn me off?"

He laughed then. "Oh, look, Bill. This is America. I know you're not American from your accent. British . . . ?"

"Australian."

"Oh, really? Not a typically raw Australian accent in my estimation. Have you – ?"

"Anyway," I cut in, "this is America. So?"

"Well, it's a democracy, Land of the Free, and all. Folks do what they want. You and Betty have plans, no one can stop you. Just friendly words of experience – wisdom if you like. That's all."

"Again, thanks, but, with respect, I've seen *A Star is Born*."

He looked at me carefully while the waiter served us lunch. "What did you say you do?"

"I'm a graduate student in ecology at Merriman."

"Yes?" he said, blowing smoke. "Interesting."

"Well, to me."

"You see," he said, "it's money that's so often the essential problem. I'm assuming that neither Betty nor you is worth a fortune just yet. But if she should make a career here, well, movies do pay well for those involved in them. Sometimes indecently well."

I nodded in agreement with this piece of common knowledge.

He suddenly bent his head in a direction over my right shoulder. "Jimmie Dugan's just come in with his female lead."

I glanced back. "Jimmie Dugan?"

"Dugan. You know, he got an Academy nomination last year for 'The Crystal Cave' – with Pearl O'Connor as his co-star."

"Didn't see it I'm afraid. Never heard of him, either. Or her."

We eventually had a good lunch. I saw no one I recognized as an actor. Later, Irve picked Betty up from Hammersmith's office and delivered us both to a minor official for a conducted tour of Cosmo Studios.

* * *

We drove back to San Francisco without any six-figure contract burning a hole in Betty's shoulder bag. But that was because – to the evident consternation of Cosmo's Vice President Hammersmith – she had asked for time to think over the offer he had grandly unveiled after viewing the results of her screen test.

The Studio was used to people wanting to consult lawyers before signing contracts. Indeed, the Studio insisted on this practice – for their own protection. What they were unprepared for was an artist of Betty's youth and relative inexperience being able to look them in the eye and say: "Thank you very much for your offer. Now I have to ask you to be patient for a few weeks. I must go back to San Francisco and talk this whole thing over carefully with several people – friends in the theatre, and also with Mr Bill Logan. We may marry before long and I have to feel sure that film commitments – if they came my way – would not spoil things for us."

I, of course, could hardly fault this attitude! But Irve Robichaud manifested severe shock.

"My God! Betty, you've left Rolph Hammersmith and his people feeling you're barely interested in their offer. Even if you – and, I suppose, Bill – decide you should sign, it doesn't leave a good taste to let them feel you're going to coldbloodedly 'assess' things when you get back to Frisco . . . and in your own sweet time."

It was then she showed the fire that could ignite her acting and, coupled with her looks and movement, would one day let her marvellous filmic qualities show through.

"Listen, Mr Robichaud, I didn't seek any of this. It was you who begged me! I'm not even sure I have any real interest in becoming a movie actress. It's the theatre I've always wanted."

"But look here, Betty – "

"No, you look, please. If I decide I want movies then I might decide to sign with Cosmo – or with someone else; or with no one. Please understand that. I won't mind too much if I have no movie career; that's how I feel at present, anyway. And, apart from my gratitude – which you have for giving me the chance of a screen test and a look at the Studio – I don't believe I owe you or Cosmo Studios anything more at this time."

For force and conviction and sincerity it was a performance that – though brief – was one of the best I would ever see her give.

* * *

Before we left for San Francisco, Irve Robichaud saw me alone for a moment. "Not my business or purpose to sway you, either of you, further, Bill." He looked at me and laughed sharply.

"As if I could! Never seen a pair of such tough-minded people under forty as you two. But one word. I've seen her screen test. Apart from her acting potential and intelligence that are both obvious, there's another quality that's more important for movies: the camera loves her. Some of the very greatest actors are not attractive on the screen. She is. If she wants a film career hers is assured. I'd literally stake my life on it."

* * *

About a month after our visit to Hollywood Betty signed a contract with Cosmo Studios. The agreement was for her to act in one picture in the next year and a half, with a studio option to have her do up to three more in the succeeding two years. The contract required her to attend a film-acting school for a total of at least six months, do a certain number of screen tests, and participate in studio publicity in connection with her acting career. Regarding the last of these requirements she stipulated certain conditions: she refused to pose for cheesecake photos, to lie about her age or family origins, to allow publication by the studio of details about her personal life, and to renounce the theatre.

She had arrived at the decision to sign mostly by herself. I had kept out of her cogitations as much as possible.

"But Bill, you must help me," she protested. "It's as much about you as me."

"It's you, Betty. Just you. Whatever we do in future can't be allowed to impinge on your decision . . . which is one of the biggest you'll ever have to make. I don't want, and can't accept, a share in it. Make it on the basis of how things seem to you. Then it'll be the truest, sincerest decision you can make. If I take part in it, you'll always be trying to factor into your thinking how my position or interests could be affected."

Cutting across her protests, I went on, "Look, whatever you do, I'm going on with what I want to do. If you decide you want us to be married anyway, that'll be great. But in fact, I'll have to think, too, just what being married to you would entail . . . so far as what I want to do is concerned. You should think the same sort of way."

"I thought you couldn't – sort of – live without me," she said. She was smiling but she sounded puzzled, maybe even hurt.

"Betty, I can't imagine anything I could do that would be better for me than marrying you. But I don't want the burden of your feeling you've sacrificed something. That would spoil it all."

Sometimes when I look back I think I can feel what I would describe as a "rueful grin" spreading over my face. How pure-principled we thought we were being, each of us, in our efforts to state our positions as selfishly as possible so as to make absolutely clear what the other would be in for. Marriage: the conundrum without an answer; the one-shot experiment in which to succeed may – on account of necessary but unforeseeable compromises – also be to fail; and to fail may sometimes be a kind of salvation for one or both.

Well, we married.

A host of Betty's friends and admirers of both sexes and more than two genders turned up. All Rudy Hauser's graduate students came, and Hauser himself, bearing a handsome cheque as a wedding gift. "My" few guests looked at her, then at me, and seemed clearly awestruck at my luck. Pete Parkes, decently muted in manner, even respectful, told me I was "one lucky son of a bitch." Betty's friends, several of them, told me much the same thing. Except for Clem Mountjoy, who said, "I admire your guts, Logan. You're an intelligent guy and, even allowing for your being smitten by Betty – as who among us males is not? – you must be well aware of the nest of problems that are lying in wait for both of you. On the other hand, if you can manage to keep the lid on things – " I wasn't sure what he meant – "a guy like you may be good for a gal like Betty. You're strong enough that she won't be able to walk over you, and you're independent and self-propelled – which is something someone marrying Betty needs to be. She's a remarkable, wonderful gal and seems bound to become famous. I hope you can handle that. I hope you both can."

* * *

We had a short happy three months, much of it filled by our work. We did manage a holiday of a week and a half on the coast near Monterey. We stayed in a modest little cabin with nothing much to commend it except seclusion and quiet. The coast was beautiful with gnarled pines and great coastal sand dunes, rocky inlets and the boil of the surf. We watched the deep red sun drop into the ocean every evening – a strange experience for me, an east coast Australian.

They were days of blue and white and red-gold sun, and bracing, briny air with the crash of the Pacific breakers always in earshot. We roamed the beaches and cliffsides in the daytime and at night lay for hours on the sand under the stars, half-awake, half-dreaming, as much in love as we could ever be.

We were three and a half months married when Cosmo called Betty to come to their film school for six weeks. I drove down with her in Maria's car. Two days after I got back to San Francisco she called from Hollywood to tell me she had been tested for the female lead in a film the Studio was considering that would be based on the novel *Tono-Bungay* by H.G. Wells. "It's going to be shot on location in England, which is the true setting for the story – if it comes off. Of course, probably nothing will happen. Most scripts are not proceeded with. They may be optioned and put on the shelf for a while, and that's the end of them. I'll probably be back when this term of the film school is over – as we thought."

Yes, well, anyway, I knew from the excitement in her voice she was hoping against hope that they would decide to use *Tono-Bungay* and that her test would win her the lead.

One week later she called again. "The thing seems to be on, Bill. They're giving me the part, and the shooting will take three months. We have to leave for London in ten days. Can you . . . can we handle this?"

It took me a moment to react. Then I said, "Sure, love. Marvellous. Chance of a lifetime for you. And, by the way, bloody good story, *Tono Bungay*. Favourite of mine. Make sure those clowns don't bugger it up."

* * *

And so the Cosmo team went to England to make the film and Betty went with them to play the leading female role of Beatrice, the lover of George Ponderevo, hero and narrator of *Tono-Bungay*.

And I stayed in San Francisco, completing my Hellhole Springs study, thinking about my professional future, and about what our life might be like now that this bombshell of an actual film career had burst, and beginning to know what it was like to be separated for the first of many, so many, times from the new wife I so loved and admired.

Chapter 16

She was supposed to be away for only three months but *Tono-Bungay* was a co-production with a British film company and there were delays caused by stoppages among unionized technicians of the British members of the film crew. Betty wrote scathingly of these delays, declaring that left- wing political slogans charging poor working conditions for British technicians had no basis in fact, and were probably being used as excuses for laziness and greed. I sympathised with her frustration, but suggested it could be advisable for her to know more whereof she complained. I mentioned how poorly the salaries of British technicians might compare with those of their American counterparts. She tartly responded that, "Maybe the Americans are a lot more skilled." At this, I reminded her that Hollywood was still happily absorbing a post-War tide of British film-makers, which indicated no doubts about their competence in comparison with the homegrown products.

Our letters, written on an almost daily basis, criss-crossed each other's, and were as near to being substitutes for normal daily conversation – in addition to affectionate caresses and murmurings of sympathy – as we could manage to make them. But regardless of my natural concern for Betty and the ache of her absence I realised with a shock how much there was that we didn't know of each other. Thus Betty, a declared Democrat in politics, apparently was readily able to ascribe motivations – that had little basis in fact – to those who seriously irritated her, and she showed scant interest in the reasons for their actions. As an Australian, with a preference for liberalism with a dash of libertarianism, it was becoming obvious to me that many differences about political and social questions might occur between us. I hoped they would be surmountable. It was unsettling to discover we had so confidently assumed that our passion, our reciprocal tendernesses and emotional affinities, could function as almost the sole basis of our bond. In fact, it was the very stuff of movies that we had already come so far without the need to test our mutual trust, tolerance, perseverance and compassion. I felt the ingredients were there all right, but I also wished that we could be living together, looking each other in the face, hearing each other's voices while we learned how to put ourselves to the test in a non-disastrous way.

*   *   *

One day a letter came in which Betty explained that the labour disputes, combined with a stretch of wet, dull weather – unseasonable for the part of south-east England where shooting was going on – would extend production time by at least two months.

In the actual, intolerable, six months which eventually elapsed, an impecunious graduate student – whose new wife was a continent and an ocean distant – did the only useful things that seemed open to him. He pushed his studies to completion and hastily wrote and submitted his research thesis, for which he was subsequently awarded a Ph.D.

*   *   *

For months, Rudy Hauser had been writing around the world enquiring about post-doctoral opportunities for me. Letters had gone to Britain, Germany, Australia, Africa, South America, as well as to many places in North America. Because of the generosity of his references and the trust accorded him by other ecologists, half a dozen opportunities had come my way. I should have long since outlined my ideas to him. But a residue of my provincial diffidence remained. I had been pointlessly timid, not wanting to put forward too many requests, or ones that suggested inflated ambitions – which might irritate him – even arouse his scorn. So I had contented myself with waiting and seeing, in the hope that something would just turn up that would give some prospect of a future move in the direction I now believed I should follow. Finally, I sidled into Rudy's office one rainy afternoon. He looked up absently from the manuscript he was proofreading. He didn't ask me what I wanted, but simply said, "Ah, Bill, glad you dropped by. I know you aren't too crazy about some of the things that are on offer, but I guess you better look hard at the last two. One's a job at University of Queensland, as a lecturer – assistant prof equivalent. The intellectual climate might not be quite as dynamic as you'd like, but you could do something yourself to contribute to the betterment of that, I'd hope. And I know from being there for a while five years ago that Queensland is a terrific place for an ecologist. I mean, where else are you going to get red kangaroos in savannah ecosystems, and tropical birds and fishes, and rain forest communities, and The Great Barrier Reef? Yeah, that could be pretty great!"

I nodded silently, and he looked at me curiously. "That doesn't fire you? Well, of course I can see that being married to a movie actress who lives in America could make Australia's remoteness something of a difficulty – even if you are an Australian." He gave a short, uncertain laugh. "Is that it?"

"In part," I said. "But there's more to it than that."

"Oh. Well, if you feel a need for more general experience before committing yourself to a lifetime academic career I can sympathise with that. We could try spots in Florida or Georgia – both with terrific ecologists. I bet Joe Higgins at Gainsville could give you a post-doc place there – maybe there'd even be an assistant prof slot if you were interested. I can – "

"Actually," I said, "I'm hoping for something else." I waited while Rudy watched me carefully.

"Name it, Bill," he said at last. "Whatever I can do I will."

"All the things you're talking about sound great," I said. "I'm very grateful for your efforts. But at this time what I'd like is not so much to teach or settle into one kind of investigation. There's just so much ecology going on around the world. I want a chance to work on something for a while and then move on. I'd like to solve problems in ecology, because I somehow think I'd be best at that – I mean as a problem solver."

"I'm not certain I follow you properly," he said, drumming his fingers a bit impatiently.

"Well, you know the wildlife biologists and the fishery biologists are dealing with questions of sustained yield and population regulation and management . . . and even conservation. These are the sorts of things I'd like to be dealing with – wherever they appear around the world, and in the setting of their various ecosystems."

Hauser nodded, now. "I follow, Bill. And can tell you something that could maybe be a solution for you. An Institute of Ecology and Conservation is going to come into being soon. It will be funded and administered by a consortium of Merriman and two other Californian universities, plus Federal and State fisheries and wildlife authorities. The philosophy underlying its establishment comes from an increasing concern for conservation of animals and plants – and indeed, ecosystems – on a planetary scale. And the perceived difficulties of getting things done in a context of real science as opposed to the polemics of special interest groups."

"Where will it be based?" I asked.

"Here, there – everywhere. Administratively, here in San Francisco. But the labs and offices and research facilities will be all over in the States, and also in some other countries. There will be only a few Institute Members – permanent staff working as ecological researchers. But it will also have Visiting Fellows and there'll be a few positions for Junior Fellows on three-year contracts. A very few of these may eventually become permanent Institute Members. Does all this interest you?"

"Tremendously," I said. "Who will be the director?"

"Yes, well, this is the point. I will be. I'm taking it on, with Merriman's approval, for a trial period of three years. Knowing you and your work, and now, a bit better, what you really want to do, I reckon we could get you a Junior Fellowship, if that was at all – "

"Say no more," I said. "It sounds too good to be true."

Chapter 17

Tono-Bungay, the film, became a critical and box-office success, and Betty made a name overnight for her portrayal of the romantically sexy Beatrice, the enigmatic female for whom the hero, George Ponderevo, would have considered the world well lost . . . if only he had not lost Beatrice herself.

I met Betty at the airport on her return to San Francisco. We embraced, and felt moved all right, but something said that things were other than they had been. In my arms was my wife with whom I had cohabited for only two of the eight months since we married. And she had come back to me different. She looked different; her hair was cut and set in a short style, but of the 1890s – that of the film character of Beatrice. The filmically arresting aspects of her features were subtly enhanced. She wore stylish makeup, not a lot but with great flare. Before, she had worn practically no makeup except on the stage. She was dressed simply, but in high fashion, in a dark, immaculately cut suit that was not tight-fitting but somehow gave the impression that it was *all* she was wearing. High heels – something I had never before seen her wear – made her walk seem less panther-like and more like that of a fashion model. Her speech intonation – formerly American, native Californian – now had a trace of an English upper class drawl. She smelled different; the sweet, clean and healthy female odour was still there, but it was overlain by a refined, but not faint, perfume.

Before she had left San Francisco, Betty had been a striking figure. The strikingness had, of course, been heightened by effort and art when she was on the stage, to make it more readily transmissible to the audience. But she was striking now in an airport crowd, and from fifty yards away. Few would have doubted if I had explained that I was in the company of a movie star. She had that presence.

She sensed what I was experiencing. As I held her away from me a little, to look at her after our initial embrace, she said, "Darling, it's me. Just me."

I kissed her again and wondered whether the six months of our separation would prove to be more than just a damnably unfortunate interruption to the furtherance of our life and love. And she sensed that, too, and said, "It'll be okay, Bill. It's over now. Things will be back to normal in no time. And whatever you think you're seeing, I can tell you it's just the same gal as before. And hellishly relieved to be back."

But I wondered if all that could be quite true.

* * *

In any case, it was not over. Within the week she was called by Cosmo Studios to do some dubbing on **Tono-Bungay**, and for publicity stills and lots of other stuff related to advertising and marketing.

We made love profoundly before she went, and again when she returned ten days later. But by then she had been paid forty thousand dollars for her acting, and was insisting we get ourselves out of our tiny apartment into something reasonable. And looking for another place took a number of days, and kept us both somewhat off balance with each other, at a time when we really needed quiet and seclusion.

* * *

And then Rudy Hauser informed me, as the newest and youngest Junior Fellow of the Institute of Ecology and Conservation (IEC), that I was being asked to begin my career by going to Wyoming for a close look at coyotes as predators on sheep. I had two weeks to get ready.

Chapter 18

Order: Carnivora
Name: Coyote, *Canis latrans*
Description: A medium-sized dog (average weight 13 kilograms)
Habits: Intelligent, cautious, tends to be solitary as adult when not mating or rearing young. Family pack consists of mated pair plus pups. Occasionally larger packs thought to include closely related animals, e.g. precocious litter members. Both parents raise young; dog guards young and brings food to bitch and pups. Growing pups playful, like domestic dog pups. Some adults show affection to each other; play and high spirits seen. Live up to 15 years in captivity, much less in wild.
Feeding: Coyotes kill live prey, eat carrion and plants. Stomach contents show:–

cottontail rabbits, hares	33%
carrion	25%
mice, ground squirrels	18%
domestic mammals and birds	14%
deer, antelope	4%
nesting birds	2%

plants	2%
insects	1%
total	100%

These were the sort of published facts and data about coyotes that were of interest to me as I churned furiously through the relevant literature, scribbling copious notes.

There was much more: their remarkable speed (up to forty-three miles per hour), strong swimming ability, migrations and long travels (up to four hundred miles from the location of their original sighting), methods of hunting, enemies (wolves, cougars, bears), habitat, reproductive behaviour.

Soon I had an indispensable handful of information about these great animals, whose original range had extended from the northwest of North America to Central America, from well-watered coasts, to high mountains, to deserts. In the research library, I had already developed enormous interest, almost an affection, for this resourceful and tenacious canine. Yet I knew my task carried seeds of potential controversy that might take me away from problems of basic biology into an area of conflict.

It was the fourth item in the diet of the coyote – the 14% of its food comprised of domestic mammals and birds – that underlay IEC's interest in this species at this particular point in its history. For those mammals were lambs and calves – and doubtless a few pet dogs and cats – and the birds were poultry. Attacks on animals that humans believe to be their property was all it took – has ever taken – to convert what, in other times and places, have been considered intelligent, beautiful and even mythical creatures, into cunning and deadly vermin. Such had been the fate of the coyote.

Coyotes had bounties on their heads in 1825, two years after their first scientific description as *Canis latrans*. Now, concern by active conservationists had caused money to flow into the coffers of IEC for an evaluative study that just might result at last in a measure of protection for coyotes in the west and southwest where men had shot, snared, trapped and poisoned them for well over a century.

* * *

I was greeted in Laramie airport by Todd Hennessy, a professor and wildlife biologist who, as a member of the far flung "faculty" of the IEC, had been co-opted by Rudy Hauser to give me some initial help in starting my project.

Hennessy was inclined to view me as a raw recruit. He bluntly said he couldn't understand why Hauser would send me – experienced in freshwater biology rather than wildlife – to study a mammal of almost emblematic significance to wildlife conservationists. "Happy to meet you, Bill, and welcome to this part of the States and all that," he said, "but I want to warn you that you may have considerable difficulty in settling into field work on a mammal as smart and elusive as this in areas of sheep rangelands. I mean, we've got graduate students, some of whom are quite experienced with local mammals, including coyotes. And really, they are the people who ought to be – "

"I'm here," I said, "because I'm a member – admittedly a very junior member – of IEC, and because Rudy Hauser requested me to do this. You're a faculty member of IEC, Dr Hennessy. If you don't think I can handle this thing I'd appreciate it if you'd take it up with Dr Hauser immediately."

Hennessy stared at me. He dragged his fingers through a shock of blue-black hair and his dark blue eyes, in craggy Irish-American features, looked troubled. "Look, Bill, let's don't get excited. You're welcome enough as a person. It's just that I'm concerned about this project. It has to do, maybe, with the survival and future welfare in this region of a species that many biologists and conservationists greatly admire. At the same time, the sheep ranchers are an outspoken group. We'll maybe have to go up against them eventually in a pretty direct way. Have you had the sort of experience that will let you do that?"

"Why don't you ask Dr Hauser? He sent me. Presumably he thinks I can handle whatever comes up."

"Yeah, well I guess I won't do that. It's just that you're new to this sort of thing . . . and you're new to this part of the States. People here can be very cordial and welcoming. They can also play hardball in a way that may shock you, if they feel their interests are threatened."

"Perhaps being new might be good," I countered. "At least I won't be trying to second guess everyone, or have a preconception about how everything will turn out."

"Okay, Bill," Hennessy said, with a thin smile. "Point well taken. However, there are some tough hombres hereabouts who – "

"If you're thinking I have any idea I can beat them to the draw or something, you're wrong," I said. "But I'm otherwise not going to worry about them just now."

Hennessy looked at me appraisingly, then nodded. "Okay," he said at last. "Now why don't we just get ourselves a beer and find out who we are."

* * *

I spent three days in Laramie with Hennessy, meeting some of his colleagues, giving talks on my work at Hellhole Springs and in Tasmania, and outlining the global tasks of IEC as envisioned by its founder and director, Rudy Hauser.

I had long discussions with Hennessy about the terrain I would be entering, how to prepare for it, the location of the best coyote sites and where the coyotes and sheep were alleged to be in maximum contact.

"You have to be careful to stay within these boundaries," he said, indicating lines on a 1:25,000 topographic map and referring to a second map that showed streams and vegetation. "We've contacted all the sheep ranchers in this area, and we hope they won't be too actively opposed to what you're going to do." He laughed a little sourly. "Of course, it won't do any harm, at the moment, if most of them think that what you'll find is certain to support their notions."

Chapter 19

In a few days I journeyed to an extensive rangeland and was conducted to a shepherd's cabin that I would occupy for the next six months, travelling daily by Jeep to watch coyotes at work and at play. It was a splendid, glowing prairie, rimmed by mountains in the near distance that were topped with a silver dusting of snow.

Several wildlife rangers took me to some of the best places to observe coyotes and to establish the trapping stations and techniques that I would be using. Mort MacDonald, the most senior and experienced ranger, was a leathery fifty-year-old from New Mexico with wary gray eyes glinting from between lids closed to slits by two generations of desert sunlight. He knew the ways of all the local wildlife and he stayed with me until he was sure I had seen most of the aspects of coyote behaviour that were on offer at that time of the year. He also took great pains to ensure that I could operate the traps and use the guns that fired anaesthetic darts with which we would capture coyotes alive. We camped out for several nights to get me used to looking after myself in this region.

Yet, regardless of his helpfulness, MacDonald seemed even more doubtful than Hennessy that I would be able to manage effectively. He was in fact caustic, if not quite towards me, then certainly towards those who had got me to Wyoming.

"They have got to be crazy," he growled, as we drove out to an observation point, "and that's being kind to them. They regard guys like me, who've watched coyotes since we was kids of ten, as less capable than young fellers like you – who's probably never seen one – just because we ain't been to college. And pardon me, but why you? A foreigner. Ain't we got enough smart college kids of our own?"

This was a less polite version of what Todd Hennesy had said.

He looked at me sourly from the driver's seat of the Jeep, then spat out of the window. "No offense, I guess," he muttered, "but the way things get done riles me."

I let this pass. Or most of it. I just said, "I can only tell you what I know. I belong to the Institute of Ecology and Conservation, and the director, Rudy Hauser, asked me to do this study. I have no idea why someone else isn't being asked to do it. So don't hold it against me."

"Sure," he grunted. "Sorry, Bill. I guess you're just taking orders like the rest of us."

This wasn't the case, because Rudy Hauser treated me as a colleague and would certainly have listened if I'd wanted to decline the project. But MacDonald's impression suited me for present purposes. I wanted him as an ally, not an enemy.

"What does concern me," I said, "is what the sheep ranchers are like."

"Like? What do you mean, 'like'?"

"I mean, will they be hostile about what I'll be doing? And if they are, will they nevertheless entertain an appeal by me for help? And will they give honest answers to questions I might need to ask?"

"Yeah. Well, as you say, their attitudes could be a mite tetchy. To them, coyotes is just a bunch of killers. But they are, most of them anyways, decent and honest guys. A young college man like you might think they're just country hicks – I'm a country hick m'self – but they wouldn't deliberately give you a bum steer. They'll tell you the truth, sure. Anyways, the truth as they see her. I mean what they believe. O' course it could be they see things a mite different from you."

"Could I trust them to report honestly the frequency of attacks on sheep by coyotes that they witness, or of cases when they can reasonably infer that coyotes have killed or injured sheep? I mean, would they inflate numbers to fit their own beliefs?"

"Look, son, I've already told you. These are basically honest guys. They won't – "

"Let me be blunt," I said. "What I mean is, are all their range hands, shepherds – or sheepboys, or whatever you call them – going to be as honest and objective as their employers are? Or will they tend to slant their reports towards what they believe the ranchers would like?"

Mort MacDonald waited quite a long time before replying. Then, "You don't give up easy, son, do you?" he said.

"It's simple. I'm going to do a hell of a lot of work on this thing. I don't want it vitiated from the outset because I can't trust some people. Whatever they may perceive as the merit of their personal reasons for not telling me the truth."

He stared at me again. "What's 'vitiated'?"

"Okay, spoiled, then."

"Shit, Bill, you'd just best ask the ranchers to tell their boys to give you the straight goods. That is, if you think you're up to telling them." He smiled thinly. "D'you reckon you're up to that? Some of these ranchers can be ornery old bastards."

"Oh, I'll tell them all right. If you'll introduce me, to soften it a bit."

He made a dry sound that might have been a chuckle. "Well, okay, son. You know I think you may be a kind of a cross-grained young feller that might not go down too bad with some of the crustier old guys hereabouts. I maybe could get to tolerate you m'self after a bit."

"Good," I said. "Glad to hear it. Let's go and meet some sheep ranchers and their . . . sheepboys."

* * *

At the end of three weeks I had met and talked with the thirty or so ranchers and their range hands. I was not sure I trusted all of them, but had to. I intended to be open and honest with them, which I hoped would build trust and reciprocal openness. I gave out questionnaires asking them to record observed attacks or harassments of sheep by coyotes, to try to estimate size, age and sex of predators and prey, to note attacks by other predators such as cougars, or even hawks and eagles where lambs were concerned. They were also asked to record time of day, weather conditions, locations of attacks and the proximity and numbers of other sheep. And I wanted to know the sizes of flocks and their main day to day whereabouts on the range.

My own programme would include watching and sometimes movie-filming the behaviour of coyotes, including their attack techniques whenever possible. Many of these observations had to be made in the hours of dawn or sunset, because of the generally nocturnal habits of coyotes. But, initially, the core of my investigation was to estimate coyote numbers, which I did by two methods. The first was to spend several weeks with MacDonald's rangers capturing coyotes with traps and dart guns, weighing and tagging the animals and releasing them, all in a twenty square mile area. Three weeks later, with the cooper-ation of some ranchers and MacDonald's staff, a great effort was made to catch coyotes again in the same area, carefully noting how many previously tagged animals were recaptured. This was a version of the "mark-recapture method" of estimating population numbers. To supplement this, and attempt to check it, I hired a pilot and light plane with some of the funds available through IEC, and flew transects over the area at low, even altitude, counting all the coyotes I could see in different map areas. These techniques – neither of them resulting in more than good approximations – gave me population values that, as Mort MacDonald muttered, "Seem to be in the right ball park, son. That's about all."

"Fine," I said. "For this sort of work we can call this accurate."

I could have used the same low flying to count sheep, and with a greater chance of accuracy than for the clever and evasive coyotes. Instead, I made sure all the ranchers knew I was flying, asking their pardon in advance for any disturbance to their flocks, and assumed they'd realise I might also be counting their sheep from above, so that they'd better be truthful about the numbers.

*   *   *

In fact, many of the ranchers were suspicious of what I was doing. A few expressed themselves forcefully.

One said, "Listen, son, your so-called study's a lotta bullshit! I tell you the goddam coyotes are eating too many sheep – one goddam sheep's a sheep too many! Now what I'm telling you's good enough. You can tell your bosses – "

Another said, "We ain't in this business for a joke. Not that we got anything more aginst coyotes than we'd have aginst any other wild, savage, bloodthirsty, mangy critter. But we – "

Most of them were more restrained, but firm in their beliefs that no investigation, however honestly and thoroughly performed, was likely to change their minds. But on the whole they were cooperative and, I believed, honest, if not always reasonable.

I felt fortunate that it wasn't going to be my task to change their minds if necessary. All I was required to do was make a report on the situation as I found it. The fight to convince or enforce – if needed – would pass to other hands, to the IEC, to conservation organizations, to State wildlife authorities, perhaps to politicians and to the State government. I contented myself with reasoning that none of these agencies could move with legitimacy, in any direction, without consideration of the report I would furnish.

For me, the real interest was in the coyotes and other wildlife, the flocks of sheep, the sweep of the rangelands, the clear, pure air, brilliant light and cornflower blue of the daytime sky, the nighttime's glittering constellations. These were what I took away with me that live still . . .

* * *

I needed to know how far the coyotes roamed – the extent of their home ranges, but also the distances that more restless individuals might undertake in their search for prey or mates, or for other unknown reasons. Small collar-mounted radio transmitters, capable of sending signals that could pinpoint the position of a coyote, even when it was well beyond sight and sound, were just then coming into general use in wildlife studies. Through Todd Hennessey I was able to obtain the necessary receiving apparatus and ten radio-collars which we fitted to anaesthetized adult coyotes.

The radio signals soon told us that coyotes from three different groups maintained hunting territories averaging a dozen miles in diameter around their dens and that they usually worked along regular trails. But two of our collared animals – both large males – travelled so far that they soon exceeded our ability to track them. One was shot by a rancher about two hundred and fifty miles from where we had collared him two weeks earlier.

Several of the many females we watched from cover produced litters and reared the young. The pups were playful like so many other juvenile canids, full of games and mock-squabbles, endearingly inventive in their play. Even the adults were playful and affectionate, tossing sticks in the air and carrying bones over distances. Regardless of whether I felt that Betty's experience at Hellhole Springs would eventually prove to have been of benefit to her – and to me, I was reasonably certain that she would have loved working with and watching the coyotes. I made a mental note that it would be vital I assume at least parts of the work I did would excite and interest her.

The killing behaviour of coyotes varied according to prey size. Sometimes the hunting by a family pack was preceded by the coyotes sitting in a circle emitting little yaps. With small prey – rodents – they stalked carefully, patiently, in the manner of foxes, leaping up to pounce down on the prey with four stiff legs. For larger prey – deer and antelope, but usually sheep – there were different tactics. In keeping with their high intelligence, the coyotes worked as a team, outflanking the prey, driving them where they could be cornered, taking turns in leading the pursuit. They were very fast. Not quite as fast as deer; we only saw them catch deer a few times. But their speed and agility were impressive; they were capable of great leaps. Foxes, though pursued, could usually evade them, because they too were very fast, and being much smaller, were even more agile. Most of the larger prey were sheep, of course, and they were not very difficult for a group to kill.

I realised that for a farmer concerned about the survival of his animals, coyotes might be hard not to hate. But for a biologist with a general admiration and affection for mammals and birds they were irresistible.

* * *

It was in this project that I knew at last, for sure and forever, that I was doing the right work . . . the right sort of science. And it was the time when the debate between facts and meaning, between the economic value of organisms and the value of lives as being – the fragile protoplasmic skin that coats the planet – became more central to me than ever before.

Now I was feeling able to argue that ecology and conservation were not merely the domain of impractical urban neurotics who had lost their focus on the "real" problems: problems like economics, life and health, societal and religious values. For to me it had now become imperative to acknowledge the rights of other creatures to survive, not necessarily as individual organisms – *this* tree, *that* bear – but as forms, as species, as populations.

Now I felt certain that their conservation – the conservation of organisms, of ecosystems – was, like history, all we ever really have of the other beings of this world. In fact organisms and ecosystems *are* history.

*   *   *

Before leaving the region I travelled south to the border of Utah with Arizona, where I managed a brief visit to Monument Valley – venue of all those John Wayne westerns – where a gruff, kindly Navajo guided me on a two-day horseback expedition among the buttes and mesas and monoliths, and we camped under the stars. And I knew for the first time that we are made from stars; the stuff of stars is already in us – and in the cactus and coyotes and sheep and ranchers – and the cities and city-dwellers. Stars are us – body, mind and being. Soul, if you like.

My Navajo guide had, of course, long understood this.

Chapter 20

I did my best to visualize the coyote as an organism with its place in an ecosystem. But I found there were too many players in that game, too many environmental unknowables. It was all too much for any study that would take less than five years. I settled for thinking of the role of coyotes in the food web.

Coyotes ate rats, several species of mice, cottontails and hares . . . in addition to sheep. Bobcats and badgers and foxes had similar diets, and hawks and eagles ate some of the same animals. A degree of competition between coyotes and other predators was therefore certain, though its intensity was unclear. As far as I could make out, however, direct aggressive interaction between the competing predator species was rare or non-existent. And all the animals commonly preyed on by coyotes were herbivores.

Predators, which are by definition carnivores, are of necessity comparatively few in numbers, or biomass, in all the food webs and ecosystems in which they occur. That is simply because they are at the top end of a food chain or, if you prefer, sit at the apex of a pyramid of organic matter or energy. There is inevitable metabolic "wastage" of organic matter as you move along the food chain or up the levels of the pyramid. As a crude approximation, the animals eaten by coyotes, and that subsist mainly on grasses, will have about six to ten times the total biomass of the coyotes. As a class of creatures often discriminated against by humans because of resentment for the damage they can reputedly inflict on domesticated flocks and herds, or because we have personal fear of them, predatory animals are particularly at risk. There is prejudice against them, often for little reason, and their numbers are low to begin with.

Sheep also eat many of the same grasses and other vegetation as the animals coyotes have always eaten. Coyotes eat some sheep; coyotes also eat many of the small wild animals that compete with sheep for plant food. So the conundrum to be answered by sheep ranchers – but which few of them even acknowledged, let alone attempted to respond to – was this: on balance, do the coyotes do more harm or good for the sheep population in terms of the carrying capacity of the rangeland?

In attempting an answer I had contrived to estimate the numbers of coyotes on the range and also the number that attacked or killed sheep.

In ranchers' terms, coyotes often killed sheep viciously and wantonly. But prejudice and emotion aside, viewed objectively, the actual number killed by coyotes as a percentage of the stocks was low – at least in this part of the range. Of course, lambs were killed in disproportionate numbers – which always aroused the ranchers to full fury. But on close questioning, most of them reluctantly admitted that the herds were not in danger of serious immediate reduction from this cause. And it rapidly became obvious that the coyotes must be living almost wholly on the readily available smaller animals.

My report, when it was finally made public several months after I had finished the study and left the region, concluded as follows:

It cannot be denied that coyotes eat up to 0.2% per year of the sheep in the representative area. Most of this 0.2% consists of lambs. Apart from its nuisance value, and the unfortunate impression it leaves in the minds of those who witness it, this rate of kill does not appear to have significant effect on the economics of sheep ranching. Moreover, given the experimental data on food requirements of coyotes (Smithers, 1947), it seems clear that the sheep killed by coyotes can comprise only a very minor fraction of their total food needs. It must be concluded that the overwhelming proportion (98%) of the food they consume is derived from predation on the herbivorous small mammals of this ecosystem. These mammals are readily and continuously available and represent a very large total biomass (Carpenter, 1951).

Young lambs are protected by ewes. Coyotes will not usually pursue lambs unless they are particularly vulnerable – i.e., injured, ill or separated from their mothers.

If anything, the loss of coyotes through increased destructive measures against them would likely lead to a substantial increase in the small wild mammals of this ecosystem and therefore an increase in competition between them and sheep for plant food – principally native grasses. It can be estimated that the result could be a marked – possibly even disastrous – reduction in the sheep-carrying capacity of this rangeland.

I was far away when this report appeared, back in San Francisco. A couple of months later I got a note from Todd Hennessy:

It's felt generally by conservationists here that you've done a sterling service. There's talk of a memorial to you, to be erected outside the University's Biology building, as a sort of patron saint of the movement!

Joking aside, everyone thinks you did a fine and necessary job. Forgive me and any others for doubting that you could. I cannot imagine anyone doing it much better in the time that was available.

However, you were fortunate to be away from the rangelands before your report came out. I understand that lynching parties and milder but more painful penalties were under discussion. Again, joking aside, there were some pretty nasty remarks made, and we cowards are thankful it was someone from far away who wrote the report, and not one of us locals. We can absorb a certain amount of abuse that will come our way from even being associated with you. But we can also act on the report while muttering something like "Well, it wasn't us, you know. It was that . . . what was his name? . . . Logan. An Australian."

If you were here we'd have to appear more loyal to you. And we would be. Over the years we will gradually become total boosters for your findings!

Thanks again, and also from Mort MacDonald, who is one of your greatest admirers. Just don't come back here for the few years it will take for the ranchers to absorb, reflect on and finally accept what you have written. And adjust their attitudes and actions accordingly. Then it will be okay, and many of them will greet you like a long lost son. People are strange in these matters.

Anyway, official policies towards protecting coyotes will have to be reshaped now, and IEC can be very happy with what you have done.

* * *

My study on coyotes and sheep was followed by a number of similar investigations by others over a decade. Some of the other work was more extensive and done over several years. Mine was the first – wherein lay its main value.

* * *

So I had taken my first steps in becoming an applied ecologist – an ecological problem solver – and a trouble shooter for the cause of conservation.

Rudy Hauser was pleased with the coyote work, and almost immediately announced he had new problems for me that we would discuss after I'd had ten days of vacation.

In the six months of my absence, Betty and I, who had really only had about six months of living together, two since our marriage, before our work had separated us, had exchanged perhaps a dozen long letters. She was now in the position of many rising young actors. The box office returns from *Tono-Bungay* were being watched avidly by the Cosmo bosses. Its critical success was assured. Betty had sent me copies of reviews. The New York Times film critic had written that: – *"Tono-Bungay*, the film, is based with unusual faithfulness on the novel of the same name by the famous British writer H.G. Wells. Many critics have labelled *Tono-Bungay* the best novel among Wells' uneven but frequently brilliant oeuvre. The film, replete with fascinating, sharply-etched characters and quirky incidents, is acted by a cast carefully selected for their ability to portray the characters of the book with precision and force. The male lead, Philip Cranston, a British star of the Shakespearean stage, is excellent as George Ponderevo, the somewhat cynical hero who is at once a detached observer of himself and others, and a rogue genius with a partially concealed romantic nature. It is a difficult role, but Cranston manages it with aplomb, and a strange convincing edginess that gives the role peculiar and memorable authority. However, the discovery of the film is Betty McMurtry, a striking beauty who plays Beatrice, Ponderevo's lover, with a blend of detachment and passion that seem very true to the persona of Beatrice as drawn by Wells. Her work in this role is a tour de force. We may anticipate many additional triumphs from this young woman as she develops in her art."

* * *

Two weeks after my return to San Francisco, four days after our ten-day vacation in Mexico was over, Betty got a summons from Cosmo. They were planning a remake of a tragic and romantic story about a pioneering aviatrix of the early 1930s that had starred Monica Healey in an Academy Award role. They could see similarities between her looks and those of Betty. But they also felt that Betty's dark, haunting physical presence could give this part even greater power and appeal than Monica Healey had brought to it. They planned to begin shooting in three months and they needed her at the studio in three weeks.

We loved each other. But, my God! we still hardly knew each other. She would soon be gone. Her contract held her to this role, even if she had not been attracted to it. But she was. Powerfully. And, as Rudy Hauser revealed my next task, I realised it could be another half year before we saw each other again. Yes, we loved each other and desperately wanted our marriage to work. But already we were asking ourselves, and each other, what marriage?

And could the centre hold?

Chapter 21

Betty occupied the first of the six weeks before rehearsals began on her new film by going down to Hollywood to view the old Monica Healey version from Cosmo Studio's archives. She watched it several times and made many notes on Healey's performance – especially those aspects where she felt she would have to do things very differently. She was not happy about her role. She believed that remakes of films were usually less successful than the originals. And subsequent experience has proved her right.

When Betty returned to San Francisco I was able to delay departure for the new assignment Rudy Hauser had handed me until it was time for her to go back to Cosmo, and we spent every spare moment together. Most of the time when she was reading the screenplay of the new version, and studying the notes she had made on the old one, I also studied for the work I would soon be doing. But we did manage to discuss her future in pictures.

"I'm sticking with what I told Irve Robichaud," she said. "I mean, I signed a contract, but basically I still feel the stage is where I belong. And I want us to somehow manage to have a life."

"Likewise," I said, "though I think we're going to have to adjust to the fact that we may not be together as much as we'd like for a few years. After that, if you join some stage company that has its own theatre and I get weary of scooting around doing IEC's bidding, and settle down near the theatre at some academic job or something, perhaps we can get it all together..."

"Bill, have we . . . have we made some sort of horrible mistake?"

"If you mean whether we can love one another, I'd say we've made no mistake. If you mean: is it going to be difficult to put a life together, the answer seems to be yes. And – "

"And?"

"And if you mean is it worth it? I think it is. Though love and devotion don't seem to come cheap or easy in the kind of a world we live in."

*   *   *

Rudy broached the subject of my next task in his office over a cup of coffee. "You know, Bill," he said, "this first project of yours worked out pretty neatly. Maybe that was good luck. I'm not going to praise you to the skies, because it's not my style. If I pick someone to do a professional job it means, right there, that I think he's a lot more than competent. I guess you could say I anticipated you'd work something out. Anyway, if it's of any interest to you, there are now three other people of your age also engaged by the IEC as Junior Fellows at work on projects in different places. There's Trevor Williamson of Oxford working on the conservation of the Nile crocodile, Pedro Jaurez on desertification in Mexican ecosystems caused by farming operations and Patrice Lafonde – a wonderfully determined girl from Quebec – who's investigating effects on caribou of changes in the tundra and taiga from hydroelectric developments. All of them have really got their hands full, because they've got long, laborious research projects and because they're encountering various kinds of political problems, with landholders and governments involved. I honestly wonder if those projects will work out for an organization like IEC. The kind of thing you had is what we will be able to do best. In and out again within the space of a single year. And without too big a profile."

"So, you're developing a sort of modus operandi for IEC?"

"That's it. We have to be an ideas operation. As for data collection, sometimes we'll have to do a lot of it ourselves. At the other end of the spectrum, where we need great volumes of information, we'll have to rely on local people to do a lot of fact-gathering, and also draw on the experience and expertise of others. IEC personnel would then become the data analysts and conceptualists. And we'd have to trust the locals to do the follow up work of application . . . and education."

"Perhaps," I said, "you shouldn't overlook the teaching aspects. Sometimes we might be very effective just diagnosing the main needs of a problem and pointing out to local people how those needs can best be tackled."

"Yes," he said, "but I also think there's a danger that you may leave a situation thinking you've shown how a project should be done, but then find people lose control of it because they haven't properly understood or absorbed what was agreed on. Sometimes, even though they seem to have grasped what we've contributed, their later actions will show they just didn't get it. Or they've successfully concealed their incompetence or lack of discipline during the short time we've had to interact with them."

"What about graduate students?" I asked. "Shouldn't there be roles for them in these projects? And wouldn't good ones greatly expand the scope of what IEC can do?"

"Not sure. Sounds okay, but many of these projects require the competence and surehandedness of people who are already experienced researchers. Grad students often make a lot of mistakes; they're in a learning phase. Sometimes they may be very, very good. But we also owe them the assurance they have a project they can finish in a few years. If people like us fail, that's serious – but it's to be expected from time to time. But grad students mustn't be led to fail on account of the project they were given being beyond the scope of what could be reasonably expected of them."

He paused, drank his now cold coffee with a slight grimace, and said, "Glad you've got thoughts about these things, Bill. Keep getting them. We need many ideas. And in fact, before you get onto this next thing, I want you to go to Chicago and talk to biologists there about the coyote work and the experience you've gained. I mean in relation to the very things we've been talking about. People are beginning to ask what the hell IEC is, and we need to fill them in on this."

He paused to scratch the side of his head. He looked tired. "I'm finding this whole IEC thing more demanding than I could have imagined," he said. "The university has granted me a year free from teaching and administration to launch IEC's activities. I'm grateful, because it means they do appreciate that it could be a valuable thing. But I have to be careful. I've always moved around more than is good for me, but a few years at this sort of pace and I'll be ready for old age, and carpentry for a hobby – if not something even less challenging."

He grinned lopsidedly. "You and Betty have no kids yet. I love kids, and Susan and I spent a lot of time with ours when they were small. But I'm feeling guilty now, because I've turned into one of these goddamned, ambitious, career-driven academics that I used to despise, and I'm on a sort of treadmill of work and responsibility I can't seem to get off of."

I had not expected such a confession from a scientist like Rudy. I looked at him, carefully, seeing as if for the first time the lines and strain. My experience so far – and especially in America – was that the work ethic applied to scientists as to others; and in spades. So I wondered if this was the outburst of a momentarily exhausted man, or if it came from deeper levels of contemplation and self-examination.

He suddenly sat up straight and smiled. "Enough of this. You wanted range and variety in ecology and conservation, Bill, and I think we can get those things for you."

He passed a file of correspondence to me. "Read this stuff. It'll give you the background details. But to put it briefly, Myron Erlenmeyer at South Arm University in Virginia is one of the IEC senior 'faculty'. Myron has documented a problem they want investigated as a matter of urgency." He then explained.

It seemed there was a stream whose source was polluted by toxic metals – principally zinc – from a recently shut down mining operation. The polluted stream, Dixon Creek, ran from the foothills of the Blue Ridge Mountains to join the Onyx River, a fine trout stream, valued for sport and scenery, that eventually ran though Wystan, a town of 30,000. The Onyx River carried ten to twenty times more water than Dixon Creek and the level of metal pollution in the Onyx was generally considered completely insignificant. The mine wastes that polluted Dixon Creek were in toxic storage dumps from the mine's operations that had contained and confined the wastes fairly well. The levels of zinc in Dixon Creek were high in its source waters, but dropped off downstream; thirty miles from the source the concentrations of zinc were all but undetectable. It was therefore easy to understand why, with the Onyx river carrying so much more water, elevated zinc levels had never been detected in it until recently.

This was just as well, as zinc is highly toxic to fish, destroying the integrity of the gills of those exposed to it, causing death by asphyxiation. Zinc is, for a number of reasons, also poisonous to many of the invertebrate animals on which fish feed.

The present problem had arisen since the closure of the mine. After prolonged heavy rains, the dumps of mine wastes – no longer maintained or monitored – had collapsed, and huge amounts of finely divided mine wastes containing several parts per million of zinc had been carried downstream in Dixon Creek, far below the former limit of zinc influence. Much of the zinc "tailings" had now either been deposited in the bed of Dixon Creek or had been spread out over the creek's flood plain. It was assumed that the effects of these many thousands of tons of deposited tailings would remain in location for a very long time. Meanwhile, Dixon Creek was now denuded of fish for thirty miles from the source. The creek became tolerable for fish only a mile or two above its junction with Onyx River. And even in Onyx River traces of zinc were sometimes now detectable.

Many people, including anglers, conservationists and naturalists, were outraged that the mining company's waste retention had proved so feeble, and were charging that the company had a legal obligation to prevent such events from happening, and to be responsible for repairing such damage as occurred. The company's position was that the rise in zinc levels in Dixon Creek had only been "fleeting", that the present levels were so low – or high so infrequently – that damage was minimal. Since no proper survey of fish in Dixon creek had ever been carried out before the alleged toxicity event, they were able to claim that there might be many other reasons for the present scarcity of fish. And finally, they asserted that they had taken the usual precautions to secure the mine wastes and couldn't be held responsible in perpetuity.

The mining company's arguments seemed weak, but Rudy noted that although there were often measurable concentrations of zinc in the lower reaches of Dixon Creek, sometimes there were not. And only infrequently had any zinc been detected in the Onyx River. This, he said, could be a problem, because the company might argue that these levels – especially as they were not constant – were biologically insignificant.

"The problem," he said, "where you're seeking compensation in situations like this, is that courts tend to go by the levels of toxic metals revealed in quantitative chemical analysis. The courts like simple lab procedures that will produce repeatable analytical results. And though there may sometimes be high concentrations of zinc in this system after heavy rains, the concentrations may drop off pretty quickly – even though there are now mine wastes all along the bed of the stream and very heavily on its alluvial flats. So we have to take a more ecological approach to this problem. Got any ideas?"

"Give me a week," I said, "to see what I can come up with."

Chapter 22

I read with furious absorption during the next few days to sample and get my mind around a huge and boring scientific literature on the effects of compounds of the heavy metals – zinc, cadmium, copper, lead, mercury – on the survival of fish of many species. There was also a lot on the effects of these metals on the invertebrate animals of freshwaters. I mean molluscs, insects, arthropods, worms and several other types of animals. But even after wading through scores of research reports, it remained difficult if not impossible to predict what the effects of these same toxic metals might be on the animal populations of actual streams and lakes. The vast majority of the tests had been set up to produce legions of regular looking "standard" mortality data. They may have provided the workers who performed them with impressive numbers of publications based on careful experiments. But they gave almost no real clues about how organisms would survive in rivers or lakes polluted by zinc. Nor was there significant information on how fish or other organisms fared in long-term exposures to zinc in amounts that were barely detectable and did not cause death in days or weeks, but might do so over months, or might damage animals' fertility and therefore the long-term survival of populations.

It had been shown that, in Dixon Creek, zinc concentrations

did not remain constant for even twenty-four hours. They might spike to potentially lethal values after heavy rain, then rapidly decline to supposedly sub-lethal levels. Or the concentrations could remain for days at a level not lethal over twenty four hours, but that might eventually kill. Any attempt to apply data and conclusions derived from the multitudinous lab tests to interpret conditions like those in Dixon Creek was bound to be frustrating.

Sometimes whole areas of science can accumulate huge collections of data that are effective blocks to genuine progress because there is no known way to evaluate them or to apply them to situations in the real world. This appeared to be one of them – a kind of "data sink", a vast accretion of facts and figures, but little insight.

I felt flat, disappointed. We were going to have to develop our own ways of evaluating the extent and severity of the toxic effects of zinc on this system.

I wandered around San Francisco with Betty, seemingly idle, but consumed by hazy schemes on how to attack this problem in the three months that Rudy told me was the maximum time IEC could allow me to spend on it.

We went back again to Sausalito where we had a leisurely lunch at a pleasant waterside restaurant. We ate seafood on a covered deck, open at the sides to glittering water, yachts, and gulls. Again we took a late night ferry on a moonlit evening like that first one. We were still in the stages of a light and happy romance. That should have been enough. But underneath the joy and sparkle, there was fear that we might have fixed ourselves into a situation that, in the responsibilities and separations looming for both of us, would make it very hard to enjoy many of the sweet and playful times we ought to have had in abundance. We were beginning to wonder now if we would forever miss out on most of such times.

We visited Betty's theatre friends, some of whom were playing in "The Life of Galileo" by Brecht, the play I'd mentioned to Clem Mountjoy and had heard broadcast by the BBC in Britain, but had never seen. This play proposes a view of science in the development and growth of the broad culture of the West since the sixteenth century. Galileo has perhaps more claim than any other single person to being the founding father of modern physical science, and is certainly one of the very greatest scientists the world has seen. But what Bertold Brecht's play about Galileo is concerned with is his repudiation of the concept that the Earth rotates around the Sun.

The reason for the repudiation is understandable and human; he lost his nerve in the face of threats of the Inquisition. But Brecht is merciless. He depicts this story as being a failure by an intellectual giant of science – at its very origins – to bring its message and its truths, its way of viewing the world and the universe, to the everyday attention of ordinary people. According to Brecht, that failure meant that common people were denied the immense excitement and expanding world view of great science, and that this failure sowed, at the very outset, the popular mistrust and suspicion of science and scientists that has eventually become endemic. It is a huge, and perhaps a simplifying claim, but I think there is a lot to be said in its favour.

Some of the cast had seen the 1947 performances that Charles Laughton co-produced with Brecht, and starred in at Los Angeles, which they declared "sensational and disturbing."

I can describe the performance we saw as "quite good", which I suppose is to damn it with faint praise.

After the performance, we went backstage and spoke with many members of the cast, and what struck me was how many of them behaved towards Betty. Some were as they had always been. Some were intimidated by her sudden fame and tried to avoid her. A few were bold enough to hint at her "desertion of the theatre". Some of the reactions upset her. "My God! there were some childish attitudes there tonight," she said. "A lot of them were simply absurd. I used to think that, as we all came up together on the stage and were all about equally poor, there was a community feeling between us that was truly comradely, and people would go to bat for you when you were down . . . and pat you on the back if you fluked a good part."

"You're just seeing the effects," I said, "not of a small triumph of a kind many of you and your friends might have experienced, but of a quantum jump where you look like you're going to end up at the top of the movie mountain, with international fame, big contracts and fat paychecks, and a mansion in Beverly Hills. It just rocks them. Some of them, anyway."

"But Bill, I'm not sure any film could be as good as a good part in a good play. I – "

"Then," I said, a bit curtly, "you better take a long hard look at where you're going, my love. Because you can't go home again.

Well, maybe you can go home again, but it could be very difficult."
After a while she looked at me. "Are you just thinking of my
stage friends, Bill? Or is there something else?"

"Perhaps," I said, "a little bit of a feeling . . . I dunno. Of
impending doom?"

"My God! Bill, I never thought I'd – "

"Hear that from me? Look," I said, "I'm human. You're my
wife. Have been for a while. How much time have we had together
. . . will we be having together? Let's get real."

"But there's your work, too. You'll be going off again almost
immediately."

"I agree. We're both roaring around, and neither should
expect the other to stop – at least at this point in our lives. But Betty,
you'll soon be in a lifestyle of money and power and prestige that'll
require both of us to do some careful thinking. If we don't do it,
we'll find the time we do have together will get invaded by seemingly
urgent events and influences that are actually pretty irrelevant. We're
going to have to fight against that."

She felt bad. I could see. So did I.

*   *   *

I arrived in Virginia by air three days after Betty had gone to
Cosmo Studios to begin work on *Flight of Fancy*. The town of
Wystan was a further hour by bus. All I had with me apart from
clothing was the plan for the research I wanted to do. I had completed
the plan the day after Betty's departure, explained it to Rudy and got
his fairly enthusiastic approval.

At Wystan I was met by Jack Schmidt, a harried looking
little terrier of a man who was president of a local group of naturalists
and conservationists – the original whistle blowers over the release
of toxic wastes into Dixon Creek. Accompanying him were Myron
Erlenmeyer, a tall, crew-cut professor of biology from South Branch
University, and Roland Templeton, town manager of Wystan.

We ate box lunches in the manager's office while I listened
to a recitation of the problem that was really a disjointed account of
what I'd already seen in the documents provided by Rudy, which
had been largely written by Myron Erlenmeyer.

Roly Templeton was a fat, sweating man of forty with an
earnest, all-business air. He had the most to say and the least to

contribute. Stuff full of "we must save our wonderful town" and "our civic leaders are determined that whatever the cost – "

Jack Schmidt, the conservationist, wanted to say and do the right things, but lacked the professional scientific insight into the problem to contribute much that was immediately useful.

Myron Erlenmeyer, the professional biologist, was the one who best appreciated the nature of the threat to the environment posed by the degraded conditions in Dixon Creek, and his ideas and comments were of some actual use.

I listened to the three of them without saying much, resisting the temptation to tell them I had already worked out an evaluative approach. I didn't think it would be appreciated if I walked in with an instant plan. After all, it was their river and their community that were being affected. When they had told me all the things that were on their minds, it was arranged for me to see Dixon Creek and Onyx River.

Dixon Creek had once been a pretty trout stream with long open runs of water, shaded pools, riffles – all the features needed by trout for habitat, feeding and reproduction. Just below Dixon Creek's source were the ugly remains of the mine. Once the mine's prosperity had supported a small boom town – Parsonville – now a ruin. The ruin's few present inhabitants were retired miners, derelicts, and people who commuted to work elsewhere. The present Parsonville would, as Jack Schmidt quietly observed, have made a good set as a dilapidated western movie town: a dusty, dry-dirt main street, one sleazy saloon, a rundown general store with filthy, flyblown windows, a scatter of collapsing shacks and cottages.

The dumps of mine wastes, as seen in photographs taken before the mine had closed, were conical mountains – piles of fine dusty material that still contained several parts per million of zinc. Amounts like this, in the economics of the times, were considered unrecoverable. Three of the original five dumps were relatively intact, except that their surfaces were deeply scarred by great grooves and channels of erosion. A thin layer of the dump material remained to mark the locations of the two dumps that had recently collapsed, but the bulk of their material had been washed away into Dixon Creek by the flood.

I said little but looked a lot. At last Myron Erlenmeyer said, "We should let you see the place where the mine wastes settled out as the flood receded." So we drove twenty miles down the course of Dixon Creek until we came to a place where it changed from a typical trout stream to meander its way across an area of fields and soft erodible soils – a small flood plain.

We examined the field, which was covered by what was clearly an inch-thick layer of material resembling that which composed the dumps at Parsonville, and which reeked of the same smell of sulphur from the crushed remains of zinc sulphide ore.

* * *

That evening Myron Erlenmeyer took me home to dinner. He was open in expressing his feelings over the collapse of the waste dumps.

"It's clear it should be the Dixon Creek Mining Company's responsibility to fix it, but it's unlikely we can get them into court. Unless you, Bill, can come up with some approach we haven't thought of." He grinned at me in a grim manner – equal parts expectancy and challenge.

"Well," I said, "there may be a way." As I sketched my plan I fancied he was looking a bit more cheerful, losing a little of the severe expression that was accentuated by his crewcut.

"So," I concluded, "we'll need stream bottom samplers and collecting nets, a small volunteer crew – probably students – to carry out some electrofishing at selected locations in Dixon Creek, some experimental field cages to expose fish to conditions along the course of the creek and at a couple of other places, and a few people who are good freshwater naturalists to identify and quantify the invertebrate animals we find. Oh, and we also need someone to collect water samples and analyse them."

"We already have lots of water analyses."

"Yes," I said, "but you may not have them from the places where most of the biological sampling will occur."

We talked a bit more about what I had planned and he was openly smiling now, which made me feel better.

"You know, Bill," he said, "Rudy called me yesterday before you arrived, to say you're a good ideas person, and that you already had a scheme that might help us work things out. On this short acquaintance, I'm inclined to think he was right. Because your plan sounds workable to me and may be what we need to solve this for us . . . at least as far as getting a scientific evaluation together that a court might be persuaded by."

"Let's not be too hasty about that," I said. "There's a lot to do, and if it does pan out we might, just might, have an interesting case. Enough, anyway, to astonish the weak nerves of the mining company so they'd be prepared to foot the bill for a cleanup by way of settlement. But you know, it'd be better if we could win in court. Then what we'd done would pack far more punch for the cause of conservation generally."

Myron nodded. "Maybe so. But, anyway, what about a cleanup? How would you do that?"

I explained some strategies that appeared to interest him. "These approaches can be blended, of course," I said.

He nodded. "Yeah. I guess. I'll make sure we get the help and apparatus we'll need. The Town Manager will help. Poor Roly. He's terrified his job's on the line over what's happened."

"Could it possibly be? I mean, how could he be expected to know the company hadn't left the site of the mine as safe as the law defines? After all, as he explained it to me, this happened years before he was appointed."

"Oh, I agree. Only, he has his political enemies. If we can beat the company into submission, Roly will love it. Tomorrow I'll explain to him what you're proposing in language he can understand."

Chapter 23

Just above where Dixon Creek ran adjacent to the mine waste dumps and received drainage from them was a small headstream of the creek. This headstream was unpolluted by mine wastes. It was a surviving fragment of what the whole of Dixon Creek had been like in its pristine state before the mine had opened. Along the course of Dixon Creek, miles below the mine site but well within the pollution zone, two other small, unpolluted streams entered the creek as tributaries.

My plan was to collect samples of whatever invertebrate animals inhabited the polluted stretch of Dixon Creek between the mine and a lower reach thirty miles downstream where, as noted earlier, zinc was usually no longer detectable. The biological findings in this section of Dixon Creek would be compared to what I anticipated would be those in the "normal" conditions in the creek above and below the zone of pollution, and in the two tributary streams.

I also wanted to establish for certain that – as had been repeatedly claimed – there were few, if any, fish in the polluted zone. For this I requested electrofishing equipment and people to operate it. Lastly, to determine once and for all whether fish could survive in the polluted stretch, I asked to have cages constructed in which fish could be placed at fixed points in the pollution zone, fully exposed to stream conditions.

The labour and equipment for this work was supplied by the South Branch University's biology department through Myron Erlenmeyer, and by Jack Schmidt's naturalists. After it had all been explained carefully to him, Roly Templeton was able to see where this work would lead; he then became enormously enthusiastic and did everything possible to facilitate our labours. Our major expenses were covered by funds from his town manager's office.

* * *

It's a commonplace, and yet to many scientists it's a paradox, that we live in an age when science pervades society, but in which so many are ignorant of science, are indifferent to it, or even actively reject it. To scientists – I mean functioning researchers immersed in the problems and challenges of their profession – it can be inexplicable that so many people appear unable to "like" science. How, we are apt to wonder, can one possibly not like an undertaking that, at its best, can give you exact understanding, precise results, that can test quantitatively one contentious claim against another . . . that can lead to prediction, decision, control?

Oh, all of us know the standard bitching about physicists and sin and nuclear doomsday; about chemists and nerve gas, biologists and germ warfare. Most scientists, any more than the general public, don't like being worried, even terrified, by demonish misapplications of science. But none of that necessarily conflicts with the absorption, the elation, of getting good results, of testing an hypothesis and finding it robust – or destroying it if it is weak; of actually being able to determine *something* of the underlying causal connections between phenomena or states of being.

At Wystan we were getting ready to do some actual science on Dixon Creek and, at least for Myron Erlenmeyer, Jack Schmidt and me, and for the students and technicians who were to help us, there was tension, anticipation and excitement as the study began. By this time we had designed all our activities so they would be coordinated, timed, so that everything – with luck – should run smoothly.

It was the job of Myron and me, as the full professionals, with some help from Jack Schmidt, to direct and supervise the work, to make sure all the samples were collected from the right places, that sound technique was used to monitor the results, to detect and correct errors. Of course, a few samples got lost and had to be re-collected; the electrofishing apparatus broke down twice; we had to readjust loose lids on our fish cages.

Our preparatory arrangements, selection of stations in Dixon Creek, training of assistants, and assembling and checking of apparatus absorbed four weeks of effort. I lived in university housing at nearby Charlottesville during that time. My quarters were cramped, but because it was the summer vacation were at least quiet. I did my daily work, when it was not in the field, in Myron's laboratory at the university.

As results began to come in from analyses of stream animals, electrofishing, and from the survivorship of fish in the experimental cages, our elation grew. It would be nice to say I missed Betty, was constantly concerned about her, but I was almost completely absorbed in what was happening. "My God!" I kept saying to myself, "this thing is going to pan out!" And when I did think about Betty, during breaks in the work, I mainly wished she had been present so that she might get a better idea of the pleasure and excitement of scientists as their ideas actually led to better understanding.

Myron was as excited as I while we plotted our daily progress, charted the results, kept Roly Templeton informed, and tried to think how best to present our findings in any legal action against the mine owners. If I had been asked at that time why I did science, my answer would have been simple and honest: "I'm doing science because it's such colossal fun! Look at this great stuff we're getting!" Of course, it's the way innumerable scientists talk when their work is running at its best. Then the drudgery, the massive study, the grinding statistical analyses, the false starts, the disproved hypotheses, the mistakes, are all forgotten . . . or at least they become worth-while. This, of course, is exactly the way the legion of scientists who worked on the first atomic bombs at Los Alamos reportedly felt, as they struggled towards what many people – and many scientists – now think of as science's most disastrous moment . . .

We got Roly Templeton to a state of such pop-eyed expec-tation with our daily briefings that Myron said, "Look, Bill, we better go easy on the poor bastard. He's about a thousand pounds overweight and he's got high blood pressure. Next thing, we'll have a heart attack or stroke victim on our hands, or even a widow who wants to sue us."

* * *

In effect, we bioassayed Dixon Creek. We found that, above and below the pollution zone, the invertebrate animals were in normal abundance and variety for a stream of its type. It was the same in the two small unpolluted tributaries. At the sampling site adjacent to the mine dumps, the fauna was reduced to a few adult water beetles and water bugs. They had not passed their fully aquatic life stages there but had flown in from elsewhere. They could survive in the creek as adults because the adults of the two insect orders to which bugs and beetles belong have impermeable exoskeletons that water cannot penetrate. At the stations in the pollution zone that were progressively further downstream from the mine, the variety and abundance of animals gradually increased until, slightly more than thirty miles downstream, this fauna looked normal again. I mean by this that it resembled that in the unpolluted headstream of the creek and in its two unpolluted tributary streams.

The electrofishing simply revealed that fish – principally trout – were found in the four unpolluted locations but not in the pollution zone – except in the station at its downstream edge, where a few were found.

In the experimental cages all fish had died within two days at the site beside the waste dumps, in six days at the next site downstream, and in twelve days at the next. Only at the last station in the pollution zone had half of the fish survived after two weeks. All the fish in cages in Dixon Creek above and below the pollution zone and in the tributaries survived for two weeks.

* * *

After two months we had all our data analysed, collated, tabulated and diagrammed. With the new academic year coming up and Myron's teaching commitments looming, we worked desperately – with assistance from Jack Schmidt – to put together a report. The University Press finally printed a hundred and fifty copies, again paid for by Templeton's manager's office.

When we met him to gloat over a pile of the reports, Templeton was in a state of glee. "Now," he crowed, "maybe I've got these bastards where I've wanted them for years. You guys help me write 'em a letter that'll tell 'em what we want. Jesus, what a good feeling this is! And" – he turned to me his round, flushed and perspiring face, that was one huge grin – "am I ever grateful to you, Bill!"

"All in the day's work," I said. "They pay me to do this sort of thing. Myron and Jack are the guys you have to thank. Myron because he's put all his own research on hold to do this. And Jack because, even though he isn't a professional biologist, he's worked like a Trojan for free. That's real devotion."

Chapter 24

The letter Roly Templeton sent to the management of the Dixon River Mining Co – now stationed in a town thirty miles from Parsonville where they were operating another zinc mine – was composed by Myron, Roly and me.

The letter noted that the Company had left the Parsonville mine in a superficially tidy condition, but that under an agreement with the Wystan Mayor's office the Company, as part of the terms of its operating lease, was responsible for ensuring that the mine wastes were secured against undue damage from weather, so that the Dixon Creek environment might be safeguarded.

The letter went on to say that two of the waste dumps had collapsed and that there was now scientific proof that zinc-containing material from the dumps had seriously damaged Dixon Creek, and constituted an additional threat to the Onyx River. The letter added that methods of mitigating the present and future threats to the environment posed by the mine wastes had now been devised, as outlined in the Research Report on Dixon Creek – a copy of which accompanied the letter.

Here, I quote from the report: –

Waste deposits from zinc mines that are many centuries

old (perhaps as much as a thousand years) can be found in the south-west of England. These deposits remain toxic for the streams into which they drain. This example demonstrates the enduring nature of this form of toxic waste.

To mitigate toxic effects of zinc wastes in Dixon Creek our advisers propose that the waste dumps should be covered by a layer of impermeable, chemically inert material such as bitumen or butyl rubber. Such a treatment would shield the surfaces of the dumps from the erosional effects of precipitation, greatly reducing the dislodgment of additional material due to surface instability. Our advisers also recommend the excavation of a wide and deep drain around the entire area of those waste dumps that are still relatively intact, in addition to the remains of those which collapsed in the recent past. The drain should be filled in with a rubble of broken rock (preferably limestone) of suitable dimensions. The drain will be linked to a nearby, unused limestone quarry that offers a very large, essentially leakproof containment basin for future releases of mine wastes from the dumps, and the limestone will ensure that the zinc is rendered comparatively harmless.

In south-west England, certain grasses have, over the course of centuries, become tolerant to zinc-contaminated soils in the vicinity of old mines, which would normally be highly toxic to most surface vegetation. Such grasses can actually be found carpeting the surfaces of old waste dumps. A cover of these grasses would probably not suffice to prevent leaching of zinc from the surfaces of the Dixon Creek waste dumps, but would likely be very useful as a safeguard miles downstream over the small flood plain where the severe flooding of some years ago deposited displaced dump material over a wide area of land – land now denuded of vegetation, useless for farming, rendered highly unstable and subject to erosion. In addition to planting this area with zinc-resistant grass, the establishment of permanent shallow drains could accelerate the return of runoff water to Dixon Creek from the dump surfaces following heavy rains, thus reducing the chance for zinc from this area to be dissolved in runoff water and subsequently deposited in the Creek.

Now back to the letter which accompanied the Report:–

It is believed that the simple procedures outlined in this Report can be used to protect Dixon Creek from further damage by the mine wastes. We would be pleased to hear of any suitable alternative procedures you may wish to suggest in order to protect Dixon Creek and its immediate environment. We emphasise, however, that an effective safeguarding should include a means for shielding the waste dumps from contact with water and oxygen (i.e. rain and the atmosphere). Otherwise an unpreventable suite of well-known biogeochemical reactions will ensure a continual release into Dixon Creek of toxic and acidic sulphates of zinc and iron, and of dilute sulphuric acid (see accompanying Research Report for details of these processes).

It is, of course, not suggested that the Dixon Creek Mining Company can be held responsible for these toxic products – which will automatically occur whenever finely divided zinc ores are exposed to water and oxygen. However, the present situation would be very much less hazardous if the waste dumps had been constructed with more regard for the damaging effects of weather, as was originally expected. The exposed dump surfaces have been severely eroded, and there has been much internal deterioration caused by water penetration through the tops of the dumps. The collapse of two of the dumps was a consequence of their careless construction and lack of safguarding (i.e., absence of drains) and has led directly to the present highly polluted condition of Dixon Creek. This creek, once a notable and pristine trout stream is no longer a tolerable habitat for fish and is all but devoid of animal life for many miles downstream. This condition will be permanent unless active steps are taken to redress the situation.

Wystan Town requires Dixon Creek Mining Company to take the necessary measures to secure the remaining dumps according to the procedures mentioned in the Research Report, or some equally appropriate alternative measures that will need to be agreed to. The Dixon Creek Mining Company is also required to stabilize the deposited zinc wastes on the Dixon Creek flood plain already alluded to.

Your early response to these matters will be much appreciated.

Sincerely,
Roly Templeton
Wystan Town Manager

Chapter 25

Dixon Creek Mining Company's response was prompt. It repeated the company's earlier disclaimer of responsibility, while asserting that it had met the legal guidelines that required it to establish stable waste dumps, and that the collapse of two of them was to be regarded as an "act of God" that, like lightning, could not be prevented by accepted human safeguards.

We had anticipated a reply of this sort. "It has not, the letter continued, "been established in a manner that would satisfy the exacting requirements of our expert advisers that conditions for fish and other organisms have significantly deteriorated in Dixon Creek during recent years. Certainly, we are not aware of events that could seriously threaten conditions for living organisms in Onyx River. We are aware that the water of Dixon Creek has, for short periods, revealed levels of zinc that are high by accepted water pollution standards. But such elevated levels appear to have been of very brief duration (as little as a few hours, a day or two at most, following flood rains). For much of the time, Dixon creek zinc levels have been barely detectable."

We read all this and Myron looked at me, Jack Gilbert and finally, for quite a long time, at Roly Templeton. Then he chuckled. "My God, guys, they've played right into our hands."

"I dunno what you're celebrating," snapped Templeton. "And how the hell do I respond?"

Myron looked across at me with his eyebrows raised. After ten weeks of working together we had a close rapport. I pretty much knew what he would say.

"Go after them in court, Roly. You can probably win. They may sense that and settle." He raised his hand to the short bristles of his hair and rubbed vigorously as we watched the doubt in Roly's puffy features. "Look," he said, "I think we can take them. If we do you'll gain an enormous lot of standing. Your problems in representing yourself as a Town Manager who cares will be over. People will respect your devotion to public issues and your ability to get things done."

"And if we can't take them?"

"Come on, Roly, you know you've got to do this. It's what you got into this thing for in the first place. Don't chicken out now."

Templeton turned to me. "Okay, what d'you think, Mr outsider-who-hasn't-got-to-live-here?"

"Look," I said, "we've established without any possibility of argument, as far as I can see, that this pollution zone is deadly for fish, and for many kinds of invertebrates. We can dispose of the company's arguments about variable levels of zinc. And in court we could point out that there wasn't always a mine, that before the mine the creek was famous for its trout fishing, and that since the mine there have been no fish in the twenty-five miles of stream below the mine, and that in those twenty-five miles of stream there is always a detectable amount of zinc, even if sometimes only a trace."

Myron chimed in to say that "Even if they claim the zinc levels are usually too low to kill fish quickly, as happens in most laboratory test studies, we can argue that the long term effects of even very low levels of zinc – which is a highly toxic substance to fish – are still likely to be the reason that fish cannot inhabit those twenty-five miles of Dixon Creek."

"And," I said, "we do have the very clear evidence of the stream invertebrates. We can claim their pattern is a much truer indication of the overall effects of the zinc and its persistence than any chemical analyses."

"Agreed," said Myron. "Though we have to keep in mind that courts do like chemical values because they're always so definite and simple. More complex interpretations, especially those involving organisms, are an intellectual challenge to them." He laughed without mirth.

"Okay," I said. "But to me our case still seems overpowering. I mean, there's legal acceptance of estimates of the effects of toxic substances or pollutants that are the results of lab bioassays. What we have to make them understand is that we have examined the invertebrate animals of this creek and demonstrated that the kinds and numbers of organisms it contains powerfully reflect the severity of toxic conditions along its course." They nodded with satisfaction. I mean, even Roly. It seemed to be looking good.

* * *

I phoned Rudy Hauser to tell him what was happening and I could almost see him grinning.

"Sounds okay," he said. "I'm sure you have a solid case. If a court challenge comes up and you can meet it successfully, it will be a landmark decision for conservationists. However, you should prepare yourself for failure. Courts, especially Southern courts, are conservative, and you can be sure there will have been plenty of local support for the company's record in creating well-paid jobs. That sort of thing could colour the outcome."

"For God's sake, Rudy," I snapped, "any sane – "

"Look, Bill, I'm just trying to prepare you for the wonders of the legal system and to let you down gently in case of some snafu. What do your friends in Virginia think? What does Myron say?"

"Okay, Myron has expressed warnings along the same lines as yours."

"Myron's a smart man. He's also from the South and knows the courts there."

* * *

I left Virginia for San Francisco the day after the Dixon Creek Mining Co. was served with legal documents from the town of Wystan. The case would be heard three months later.

Chapter 26

Leaving Wystan was hard. I had formed a bond with Myron and Jack Schmidt – even with Roly Templeton who, beneath his fat and sweaty exterior, was a sincere and honest public servant.

As my plane descended towards San Francisco I was coming down myself from an elation that had gripped me for weeks. Now I wanted to give Rudy a good verbal account to go along with the printed report. I half-feared he might punch holes through our reasoning and point out many things – many obvious things – we could have done to give our results more point . . . more bite, if it came to a legal wrangle.

But his comments were few and mild. I didn't know whether he was preoccupied with other concerns or just not turned on by the study we had done. He had seemed interested enough when I'd given him telephone reports from time to time. Later I came to realise that he had been pleased with our efforts, but that, as with the coyote work, he had expected that I would do a good job. As he told me at a still later date, "Your only problem would have been if I had found it necessary to express my disapproval or disappointment with what you'd done."

* * *

As soon as I'd seen Rudy, I telephoned Betty in Hollywood.

She wasn't in, and it took me two days to reach her.

"Bill," she said, her voice husky and deep, as if she were exhausted and perhaps depressed, "Is that you? Where are you?"

"I'm home. What you could call tired but pleased. What about you?"

"Well... tired, anyway."

"Not pleased?"

"It's complicated. I need time to explain."

"When will that be?"

There was silence. I repeated the question.

"D'you think you could come down here, soon?" she asked.

"Well sure. Of course. Only, I was sort of hoping there'd be a break in the shooting, and you could come home here to San Francisco for a day or two."

"Bill, I want to. But things are pretty crazy here. The shooting schedule is nuts. The whole production's becalmed. There are technical and directorial problems you wouldn't believe. The director's losing his marbles. And . . . so is everyone. So am I. Please come down. Please! If you possibly can . . ."

* * *

I arrived in Hollywood about 4:3O p.m. on a blazing early September day. The smog and humid air of summer had been blown away by refreshing ocean breezes. It was hot, dry, clear and bright, and made you see why people had wanted to settle on the Pacific coast in the first place.

Betty lived in a nice suite of rooms in a building on the Cosmo lot. It was the studio's policy to offer such accommodation at modest rent to those among its younger and newer actors who were under contract. She was not there, because they were still shooting scenes that had begun at eight in the morning, but she had arranged for a security officer to let me in.

Around five o'clock a studio bus brought her from where she had been filming, still dressed in her aviatrix outfit for *Flight of Fancy*. Her hair was dark again – its natural colour, and longer. Not shoulder-length as it once had been, but longer than for *Tono-Bungay*. She looked slimmer than ever, the skin stretched almost tightly over her delicate high cheek bones, her handsome mouth made up dramatically, her eyebrows arched; otherwise she wore no makeup. She looked a mixture of things: dashingly beautiful, lissom, vivid, commanding, yet somehow vulnerable. She also looked tired. For the first time I was aware of tiny lines around her eyes and the corners of her mouth. They looked attractive when she smiled. But they were there. They hadn't been there before. I was looking at the face of a beautiful young woman, but it was no longer the face – nor would it ever be again – of a girl. I felt moved and sorry, alarmed and indignant.

We embraced and, to my amazement and dismay, she burst into tears. I had never seen her do that, and it had seemed it would never be part of her. Her hallmarks seemed to be determination, indomitability. Now she was sobbing the tears of any female who has lost control in the presence of one from whom she felt no need to hide her feelings, and could, for a moment, offer a shoulder to lean against.

Apart from being sorry for her, I suppose I felt flattered. Anyway, I felt unaccustomedly large and strong and confident, as well as tender and sympathetic.

"Darling," I said, "Betty dear, what is it? Don't. Hold on."

She clung to me. "Monica bloody Healey," she sobbed – half laughing, "that's what."

"The role?"

"The role's okay," she gasped. "The role's wonderful. But Healey's ghost haunts it. Russell Lambert, our director, knew Healey. He was a junior assistant director for the first version of this film. Lambert swears the last thing he wants is for me to reprise Healey's performance, of which he's quite critical. Yet the bastard's forever recalling how she did it and wondering if I should do it her way or completely differently. . . or just my way, if only we can discover what that is. God, how I've learned to loathe Monica Healey . . . and Russell Lambert."

"He's a hopeless director, then?"

"I didn't say that." She was suddenly guarded. "He's very intelligent. Very able. I think he hates me. He – "

I took hold of her shoulders, held her at arm's length, looking at her carefully.

"Betty," I said, "I can't follow this. Is it really all so bad?"

She was choking back more sobs then, but finally she stopped. "No. Of course it isn't. I'm sorry, dear. I love you." And she kissed me.

* * *

Next day she took me on the set and showed me the amazing mockups of aircraft being used in the film. Then I met Russell Lambert. He was about fifty, grey-haired, short, with an authoritative, slightly didactic manner, but not unpleasant. I thought he seemed a fairly level-headed person. While Betty was getting made up he spoke to me.

"Your wife, Bill, is undoubtedly – or will undoubtedly become – a great film actress. She has passion and sincerity. You can believe in her. She's intelligent in how she speaks her lines and she has subtlety of understanding. She grasps the understatement needed for movies. She can be very still, except for her eyes. She moves like a dream. And, of course, she photographs marvellously. After I saw her first work in *Tono-Bungay* I knew immediately she could do *Flight of Fancy*."

"Good," I said. "Glad to hear it. Can I be frank?"

"Of course . . ." He stared at me; he looked a bit defensive.

"Well, I haven't seen Betty for three months because I've been at work in the east."

"Yes, in Virginia. She told me."

"First thing she did when we met was fall apart. Came unstuck in my arms. That's not like Betty. She's usually an indomitable lass."

Lambert examined me for what seemed a long time. I found it more irritating than unsettling. He said at last, "I assume this was in relation to *Flight of Fancy*. I mean, if it's some more personal matter it's probably not appropriate to tell me about it."

"It seems to be about you," I said.

"Okay," he said. "I think I understand. But let me put it this way to you. Betty may think it's about me. But it's really about loneliness. Actors as a class – many of them – are very high-strung people. Betty may have seemed pretty calm during her stage career. But there, no matter how prominent you are, it's not so lonely. You're one of a crowd, a family almost. And you simply have to keep a hold on yourself to perform every night. But movie acting is a very exposed sort of enterprise, regardless of how famous the actors become. You may be fussed over and the centre of things in a way no stage actor is, but the intense and very personal interaction with many others of your craft just isn't there to nearly the same extent. Except maybe with the director. And that relationship can become terribly focussed and intense. And as you deliver your lines there's only a silent technical audience and a cold glass eye to watch you. Betty's only a young woman. You're her husband, but you've been away. I'm afraid, Bill, there'll be more of what you saw yesterday. Unless she quits movies. Or unless you give up the work that takes you away. And, of course, I don't advise anything like that. She might not even respect you if you didn't have a career of your own. What's even more important, you might not . . . "

"Respect myself?"

"Exactly."

"Anyway," I said, "there isn't any possibility of my giving up what I'm doing."

"Okay. Good. Then regardless of how it might seem, you may have a better chance of things working out for you than if you weren't both inclined to follow your own star."

"Hope you're right."

"You know, Bill, if I could say something really useful to you both, it might be 'avoid the movie business', she as an actress, you as her husband. I doubt it can bring you much good. I speak as a friend, as Betty may again think of me when this movie is over. My feeling is that if Betty stays in it there'll be lots more of what you're both experiencing just now. Or she'll have to be tougher, learn to endure the process. And you'll have to do the same. Both keep a sense of humour and proportion. Be tolerant and forgiving. Live and let live as far as work is concerned. The part of your lives that you really share may have to be a separate affair where you can find some common ground. Perhaps if you have kids some day . . . "

None of it seemed satisfactory. I didn't dislike Lambert but wasn't sure his advice could offer much to live on that wasn't self-evident. Tolerance, for God's sake! Said like that it sounded like thin going, tasted like ashes in the mouth.

* * *

Betty pulled herself together and next day seemed buoyant, happier about the film, not so exercised about Lambert's direction . . . and disinclined to discuss or even to speak further of the things that had vexed her a day before.

It occurred to me that Lambert might be missing the point, that even if she had shown little of it before, Betty might be treating me to my initial first-hand experience of the "artistic temperament". If so, it was a sobering revelation. Would it grow, nourished by the grovelling adulation and inflated income that are the common lot of film celebrities?

Well, I loved her. And we made love, and much of the time during my visit things were as they had been.

She introduced me to her fellow actors. The male lead, Elliot Rogers, was required to play a stiff English type, a reprise of the original male lead of the earlier version of *Flight of Fancy*. It was easy for him, because he *was* a stiff English type. Apart from that he seemed a decent fellow and, over a beer between takes, we talked of cricket and rugby and English pubs. He told me there was a keen bunch of cricketers among the many expatriate Brits in the Hollywood film colony. Three or four of them, he gave me to understand, were "really bloody good, old boy, and practised and played like buggery all through the summer, if not actually filming."

He looked at me. "D'you play yourself, old boy?"

"I used to," I said. "Played at university, but I'd be very, very rusty now."

"We have our friends play on our teams," he said. "Delighted to have you along for a game, what!"

"Well, thanks very much. I might just take you up on that."

He glanced over at Betty, who was preparing to go before the camera. "This lass of yours, old boy, terrific little actress, y'know. Going to go places. Smart little woman, too. Not all of them here are, y'know. Pure gold, this one . . . And what a looker."

I nodded. "I think so."

A bit later I watched my beloved in front of the camera, just before she prepared to climb into the cockpit of a plane mock up, in a long, extremely intimate embrace and open-mouthed kiss with Elliot Rogers. I stared, feeling, if not quite looking, equally open-mouthed in my dismay.

At that moment, long after that moment, and again, long after that, I found myself groping for those banal words of Russell Lambert. What were they again? I could see them, those words, like stupid woolly sheep straggling over some distant dim landscape, while my numbed and limping mind struggled, like some drunken shepherd, to muster them. Surely I could gather them up, those poor, hackneyed

words that were somehow still the truth. What were they? Oh, yes!
Tolerance. Live and let live.
        Sure . . .

                        *   *   *

        The shooting continued. I spent a week at Cosmo Studios
and on the whole it was fun. But Betty kept talking about how great
it would be to get back eventually to stage acting and her old San
Francisco theatre crowd. This began to sound like a mantra. Then,
as she kept it up, I decided it was starting to sound more like neurotic
chatter. I told her this at last, probably unforgivably.
        She was furious. "What the hell d'you mean?" she cried.
        "I mean," I said, feeling I must stick to my guns, "what I say.
You must realise you can't go back to a semi-amateur stage company.
The poor devils would feel crushed and patronized out of their souls
by having a current Hollywood romantic star slumming among them.
You might well get a break in the theatre on account of your rising
fame. But not with your old company. And you know that, Betty."
        We wrangled on. It didn't quite spoil things, but for the
moment our enjoyment of each other was dampened a bit. I was
annoyed for having spoken so bluntly, and confessed to myself that I
was too often short with her. She would have worked things out for
herself soon enough.

Chapter 27

Back at Merriman, copies of the correspondence between Roly Templeton and the management of the Dixon Creek Mining Co. had arrived. It was pretty clear how they would build their case, but we had to give the documents full attention. Environmental law was in its infancy, but Rudy knew a few intelligent and idealistic lawyers who had biological training and who were devoted conservationists. I had long talks with two of these people, and Myron Erlenmeyer came to Merriman to discuss everything with us in the presence of a keenly focussed lawyer by the name of Geraldine Howkins.

I knew a court case lay just down the pike and was willing to put everything into it. But mental fatigue was setting in and I knew I needed a change.

Everything was made worse by the separation from Betty.

Rudy sensed my mood. We were in his office after our third discussion with Myron and Geraldine Howkins, and he seemed nervous, drumming his fingers on his desk top as he looked at me. "Bill," he said, "at Hartland University they have a series called the H.C. Gustafson Lectures. Heard of them, by any chance?"

I shook my head.

"Well, Gustafson was a pioneer naturalist and conservationist in the late 19th Century. Remarkable man. Millionaire industrialist, later U.S. Attorney General. As a businessman he began to understand the damage that industry could wreak on natural ecosystems, and as a naturalist he determined to use his wealth and prestige to do something about it.

"He set up a fund that does several things. It supports a conservation foundation, it awards post-doctoral fellowships, and it endows a chair at Hartland. Another part of the fund supports annual lectures on the practice and philosophy of wildlife management, conservation, ecosystem preservation, and so on. Here"– he handed me a brochure – "is an outline of the lectures and their scope. Every year they have two speakers. I've been asked for nominations and I'm naming you."

"I hardly know what to say," I said cautiously. "Of course, I'm greatly flattered, but . . . this sort of thing's for seasoned veterans, surely?"

"Oh, God, of seasoned veterans, enough already! We need the voice of 'now', or even of the future. You'll do it, won't you? It'll maybe get your mind off Dixon Creek and more personal concerns for a while. Give me all your c.v. material, let me submit it with your name, and we'll see what happens."

*   *   *

It was a month later when I stepped to the podium, spread my notes before me, and took a deep breath. The audience was about two hundred and fifty, many of them Hartland faculty from various departments, including a number from the humanities.

My subject had been advertised as "The Lost Domain" – a title I lifted from the English version of Alain-Fournier's great novel. The notice also mentioned "The Conservationist Stance in a World of Constant and Rapid Environmental Change" as the essential theme, and indicated that the concepts of Community, Ecosystem and Landscape in relation to Conservation would come up for consideration.

I faced the audience, trying to be relaxed and reasonably confident, and began to speak.

"What conservationists have to do is find ways to convince their millions of doubters of the integrity and the necessity of the conservation enterprise. In other words, that conservation really is an idea whose time has come. Because for all too many, the mere mention of conservation immediately raises their hackles. Their first question, hard, direct, perhaps hostile, may be –

"What do you want to conserve?

"That's fairly easy to answer: particular species, communities of organisms, ecosystems, natural landscapes. To which their response may be: 'Oh sure, like plenty of big parklands and some of the more spectacular beaches and mountains where you can camp and swim and hike and trail-ride and ski. Sure.'

"This apparently amiable concord may run out very smartly as you insist, gently but firmly, on defining the terms species, community and ecosystem, and clarifying your use of the term 'landscape'. It might, of course, be assumed just about everyone understands what is meant by community and ecosystem. And absolutely everyone knows what you mean by landscape. Well, many people recognize those words, but many people also recognize the words 'space', 'energy', 'momentum' and 'relativity', without necessarily appreciating their real scientific meanings."

I looked at my audience. Were they at all interested? Well, at least they were not moving. Were they then in a trance or on the verge of slumber? I shrugged – mentally.

"I am going to presume," I said, "that even in an audience as learned in various departments of knowledge as this one, there may be a few for whom the terms community and ecosystem – though simple in concept – will actually have less than everyday familiarity. So bear with me."

I drew myself up and, like a preacher intoning a prayer, recited to them:–

"In the usage of ecology, a community is an assemblage of organisms occupying a particular area or region. An ecosystem is a system formed by the interactions between the organisms of a community and with their environment."

I took a deep breath. My audience gave no apparent sign they were affected by my words.

"Anyway, even when people do understand what is meant by these terms, such understanding, far from winning them over to your cause, may merely rekindle their initial suspicion and hostility. Why? Because now they know they are in for particulars rather than woolly abstractions. They sense that you are going to be specific about what you want to conserve, and its location. And as soon as they know those things, the next question will again be of the hard, direct kind: 'Why (or perhaps, why in the name of God!) d'you want to conserve *that*?'

"And then there are those who will ask a series of questions, in a seemingly reasonable way, that are actually much more difficult and troublesome for the conservationist, because however he chooses to answer, the questioners will already have thought their way to elaborate counter arguments and obfuscations. The questions go something like this:–

"Surely you will agree that civilization is built on the invention of agriculture, on the necessary *modification* of ecosystems, on the taming and exploitation of environments too harsh or threatening in their natural brute forms for humans to cope with? Hasn't it always been obvious such ecosystems and environments must be converted into states suitable for growing crops, raising livestock, mining, building cities, and into places where we can stroll confidently at our unmenaced ease – and go forth and multiply? Shouldn't we – in justifiably ranking our species as the acme of evolution's products – expect to continue to dominate what we have long referred to as the blind forces of Nature; I mean continue as we have already done, with ever-increasing confidence, since the very dawn of civilization?"

I looked out across the assembly, and raised my voice a notch.

"I put it to you: Is it not almost universally considered *The Right Thing* to modify Nature's landscapes as we have done and are still doing – now at a greater pace then ever before?

"And now for a moment, consider *landscape* – a term that is broader, more inclusive, and employed more confidently by non-ecologists than the terms community and ecosystem. Landscape is largely a modern concept, an invention of only the last few centuries. Before that, in our species' only visual records of nature – the arts – we were concerned nearly always with depiction of *people*, of men and women, gods and goddesses, heroes and heroines, nobles and peasants at war, work or play, or animals, or a few aspects of vegetation, or with great buildings – all in either hostile, or unrealistically sublime, and anyway imaginary, worlds. Where landscape was featured in such art it was nearly always as a backdrop to the central subjects – often in the form of stylised, romantically beautified parks, or sublime distant mountain ranges, or wild romantic seas. Hardly ever was mere, actual landscape – landscape without people, or even landscape that had been reconfigured by human activities – depicted as a central subject in art. Landscape was thought of as the result of the impact of Man on Nature and, to give landscapes significance, people needed to appear in them. Nature – natural ecosystems – was the raw material for human expression, and landscapes were the result. People did not measure themselves against Nature; they subdued and dominated it.

"Landscape itself, as a legitimate subject for the artist, blossomed late in the day. But I doubt that interest in landscape could have come into its own before our power to overcome it had established in us the confidence that it could simply be viewed 'for its own sake', our perception of it as a brooding and menacing presence no longer any more disconcerting than a caged tiger.

"The truly sticky question is how, in the course of human history, people came to fear and reject Nature. How did we come to believe that the only reasonable thing to do with Nature was to conquer it, change it, and use it?

"It has to have been because of technology!

"The technology of farming gave us our first power to change Nature to a degree out of all proportion to our numbers, allowed us to settle in one place, eliminated the necessity of a hunter-gatherer existence and the immediate and total dependence on the family or tribal unit. Over time we forgot how it had been to live day by day as part of Nature, and gradually became alienated from it, came at last to fear and mistrust it.

"Yet – and this is the great wonder – humankind universally dreams of, and often prays for, gardens of glory, places of unity, ripeness displaying variety yet an inner cohesiveness, places of plenty and of repose. Places that could be regarded as beautiful, and often complex, ecosystems. Places, I suppose, like the Garden of Eden.

"Lately – I speak of the last few centuries – some of us have at last re-acquired what we must long ago have misplaced: the self-assurance to confront Nature head on, full face. Was it this confidence that led a Wallace and a Darwin to look on Nature with their senses of curiosity and wonder given full flight, and that has led us, through them and others like them, to the modern science of ecology and the practices of conservation?"

I gazed out into the dim lecture room – an old room in an old building, part of a university quite old even by European standards. I thought, as my words floated out past the potted palms that flanked the speaker's podium, that a century earlier someone could have been addressing an audience of similar size and level of academic attainments in this very room, on a similar muggy Wednesday afternoon. Someone droning on like me about the discoveries of the day – perhaps about the dawning age of electrical engineering, or early ideas of thermodynamics, or even some exciting discovery of paleontology or astronomy. Now here I was, trying to lead back from the technology that was still changing everything, to more primal and obscure, but more transcendent, matters. Scientists as a species had long learned to dispense with what were commonly viewed as ultimate questions – questions of whence and why and why not. Questions of means were now almost completely dominant over taboo questions of ends.

Well, I had decided, even if it compromised my standing as a serious biologist, to raise some questions of ends.

I took a gulp of water and went on.

"It is to modern biological science that we owe – under the general rubric of ecology – not only the concepts of population dynamics and ecosystem, but also of community, succession and climax. As concepts these, and other related concepts, are accessible enough even to many non-scientists. They lack the occult formulations of mathematics and the subtleties of philosophy. Yet – and this *is* astounding – only a century ago there was hardly anyone alive who thought of living Nature in terms of such concepts. Like many of the most basic ideas of biology they are not particularly difficult to explain or to grasp, yet they have required an unaccountably long time to establish themselves in the minds of many people of good general education."

I paused, drank water, took a second deep breath. Wondered if this stuff was likely to go down at all. Plunged on.

"Let us be clear about one thing. *Wild* Nature is popularly thought of as anarchic or chaotic, but is in fact subject to rules of patterning and order, dictated by physical conditions, by the relationships and interactions between species and between and within populations, and between individual organisms as they multiply, live, grow and die.

"To understand these relationships and interactions is, of course, to understand how ecosystems function.

"Let us return to that universal image of a beneficent, welcoming Nature: a Garden of Eden. Apart from its mythical status as a region of spiritual and biological perfection, it represents a place of abundance, a biological comfort zone, a refuge for the body and the soul in a universe that is otherwise a dark unknowable place. In the Garden of Eden we should perhaps expect a splendid, shining biological locus, where self-completeness, sustenance, beauty, order, both simplicity and complexity, permanence, essential mystery, the conditions for reverence and love, all are present. We should expect the prospect of endless contemplation and adoration of the Garden. When we speak of this Garden, we are clearly employing it as a metaphor for biological Nature in all its marvels and mysteries – a Nature that, spiritually speaking, we are still part of."

Now I looked out at the audience again for a long twenty seconds. What could I make out in what seemed to me impassive faces? I turned my eyes down again to my notes.

"Anyway, I would like to assert that some time ago – let us say for the sake of argument between ten thousand and a hundred thousand years ago – most or many of the ecosystems of this world that were occupied by humans could have been held as local approximations of this Garden of Eden."

I found I was feeling really parched and began to long for a cold beer . But I went on . . .

*   *   *

Fifteen minutes later, my lecture over, I gulped down a full glass of water and looked out to face questions.

A rough-bearded man in a tweed jacket, with some sort of Camford accent, said, "But surely many of the ecosystems that were abundant in the past would have lacked what I will call the 'luxuriousness' of what is commonly thought of as a necessary condition of The Garden of Eden – I mean the sorts of condition that would make it an appealing place to spend eternity in." He chuckled, as if in mild self-satisfaction at a successful debating point he had made. And I realised with a slight start that he was wearing a clerical collar.

"Oh, look," I said, "of course they would have varied regionally, climatically, geographically, in their details. Some were jungles, some were near-deserts, savannahs, swamplands, rich coastal lands, mountain fastnesses, lakes, seas. A few were maybe icy wastelands.

"The animals in these Edens did not love one another. The lamb did not lie with the lion – at least not while alive!" I heard a few laughs. "But they were places where the human inhabitants passed their entire existences, where they knew the edible plants and animals. They didn't know the microbes science has discovered, or the chemistry of photosynthesis, or even the primitive elements of agriculture. But in a lifetime they would have been able to acquire much understanding of their environments – the ecosystems of which their environments were a part – and how to stay alive in these environments in a certain state of harmony ... or at least of balance."

Someone said, "Are you supposing that these primitives possessed a pretty profound ecological sensibility?"

"Not so much that," I said. "But inhabitants of an Eden/ecosystem would soon learn to associate a food plant's occurrence with a soil type, a shade, a particular exposure to, or shelter from, wind and water and sunlight. And would soon learn the habits and shelters of the animals they sought for food – or to avoid.

"It would be just one further step for the thoughtful among our ancestors to begin to contemplate the wholeness or, if you like, the interconnectedness, of the conditions in the ecosystems in which they lived: the rain that fell in hills or mountains and led to spring growth, and made its way to streams and lakes; the long days of summer maturations; autumn, when growth had ceased; winters, when survival would depend on rigorous adherence to rules of experience.

"And our ancestors – and the same with all remaining primitive societies in the world today – would have been concerned with forms of contemplation. These forms of contemplation would not have resembled the profoundly objective, mechanism-oriented contemplation of contemporary science, but would be more like the dreaming, unselfconscious contemplation of the sun, the moon, skies, clouds, birds, winds, storms, lightning, mountains, lakes, the sea, trees, the very big, the very small. And heat and light, cold and dark. And stars. What the remaining primitive societies of today still meditate on."

The questioner looked doubtful, remained standing as if to ask something more. Guessing what that might be, I said, "Look, we've gone through at least ten thousand years – four hundred generations – in which we have got very far away from the world of the ecosystem. We're living in times when our power to change ecosystems extends to their appearance, structure, function, to the possibility of pushing them in ways that are dictated by our self-defined needs of the moment. If we are so disposed, we have the power to totally destroy any ecosystem, with all its plants and insects and mammals and birds. I mean, if we place the value of some commodity in that ecosystem beyond the value of all other attributes. We are limited in what we can do only by money and energy . . . and the extent of our ruthlessness."

"But returning to what you said earlier about The Garden of Eden, what about the gardens in our present world? Could you say something that relates to them?" asked a pretty woman, who looked far younger than the majority of the audience.

"Well," I said, " if the Garden of Eden can be thought of as a mythical embodiment of the ecosystem, the gardens we have constructed during historical times – whether great formal affairs, or modest domestic plots – have perhaps been the products of a sort of spiritual revenge for being, as it were, excluded from the primal Garden. Maybe resulting from a kind of unacknowledged fury at being somehow disinherited. Sort of like saying that, because we are now denied the bucolic perfection of the ancient ecosystem-Garden, the gardens we make as compensation – and for . . . revenge – are in most cases ecologically simple or ecologically unstable. Most of them are simple enough so that we can manage them, maintain their ecological structures as we wish. We control the weeds which would rapidly replace them with confusing tangles of growth that resemble neither the archetypal Eden-gardens of dreams, nor the gardens-as-art we fashion in their place. Of course, most untended gardens will disintegrate over time, coming to resemble the sorts of ecosystems which preceded them, and which may still occur at their margins.

"These gardens we make will be safe, so that we need not concern ourselves over the possibility of encountering snakes or tigers – or other types of creatures that could threaten us.

"They are *ideal* gardens. We don't eat their fruits; most of them will have *no* edible fruits – fruit-growing is the province of agriculture!

"Never think of the gardens we construct as being anything more than simulacra of our long-ago Eden. We don't live in these gardens; mostly we visit them only to shape, modify or weed them. Since the complex natural gardens of which we were a part are now denied us, the gardens we make will be simple enough for us to dominate and make them entirely ours."

I sipped from another glass of water, now grown tepid and

smelling of chlorine, wishing profoundly that it was a large cold beer. I was feeling weary. But I put down the glass, took hold of the lectern, and went on.

"I'm going to say something to you, as if all of you were scientists with a wide-ranging humanism. We *need* ecosystems, not more man-made gardens but great representative chunks of the major natural ecosystems of the world – lakes, rivers, plains, rain forests, savannahs, coral reefs, rocky shores, wetlands, the oceans – and also examples of the smaller, rarer, less-acknowledged ecosystems: like thermal springs, salt marshes, small uninhabited oceanic islands. As never before in four hundred generations of our species, we need to understand and appreciate – through the remarkable science now at our disposal, with the vast history of aesthetics and spiritual wisdom now in the record, all of this available to people everywhere as never before – we need to understand and return to the contemplation of Eden."

* * *

I had been asked to talk to a group of students, and next morning I met about fifty of them.

They seemed a lively bunch, less deferential than the academics of the day before – which made me feel easier. Some of the ground my talk covered had already been presented at the Gustafson lecture, but now I went on.

"History, and the fact of our ever-increasing numbers mean we can never again expect to inhabit aspects or versions of the Garden in the complete way that our forebears did. But we must strive, nevertheless, for understanding, not only of great engineering and immense agriculture, but for appreciation of the world we evolved in as a species, a world that somehow still lives in bright images in our dreams. We must find ways to return to contemplation of that world and empower ourselves to enlighten people everywhere to pay heed again to its tempos and its laws – which are things we nowadays ignore in every hour of our waking lives.

"Conservation is our sole means of ensuring that, as the combined exaltations and terrors of technological change threaten to swamp us as creatures of feeling and spirit, we need not lose all contact with our roots, nor allow the magnitude of our hubris to lead to our disinheritance.

"We have all heard of 'The Two Cultures' – this notion of deep separation between science on one side and the arts and humanities on the other. But thousands are drawn to study science through love of the beauties and mysteries of Nature. If we learn more of the workings of ecosystems through science, will they then cease to be objects of cultural concern whose wholeness we can venerate and meditate on? I think not! Scientific understanding cannot diminish the essential sanctity of the objects and processes of Nature, because our studies will not ever quite pierce to the central core of the matter. There will always be a next level of understanding to supplant the present level. As we push ever deeper in our searching, and seem to approach the still point of the turning world, the deepest meaning of Nature will continue to elude us. Or so I believe."

A student, a freckled, sleepy looking kid of about nineteen, asked me just what I thought was of such special significance about landscapes. "I mean," he said, "millions and millions of people now live in cities, never really see the countryside, let alone primal ecosystems, and that's likely to be the case even more so as time passes. Do you seriously think this is bad for them? And why should it get worse?"

"Your question's good," I said. "But I think that for most people who really have the opportunity to experience life away from cities, they do, in increasingly large numbers, seek and find ease of spirit, and beauty, in the landscape. I think most of us know why our hearts feel comforted by the landscape. It is home. Most of our species stopped living in that home long ago. But the dream goes on."

"Okay," he said, with a slow grin starting on his face. "Actually, I agree with you. I just put the question to see what you'd say."

I laughed and many of the other students chuckled.

"But," he went on, "why does the Garden – the ecosystem – seem to have this hold on us, those who've really experienced it?"

"Well, let me put this to you as a simple question. Why do we love clean water, blue skies, white clouds, rich vegetation, golden distances, trees, rivers, lakes, the sea? It does little good to say: because of our awe of them. Or because of the sense of mystery and transcendence they inspire. Those things surely come later, as products of the cultural accretion of recorded images in prose, poetry and music.

"Take water. If we like it clean and clear and sparkling it's also less likely to contain a concealed threat to health and well-being. A blue sky signals fine weather, is insurance against bad. White clouds can also reassure – unless they are storm cumulus; but even those may signify relief from drought or heat. A great wind may refresh us after heat; sometimes, of course, it can turn dangerous for a while!

"Healthy-looking natural vegetation suggests we are in a comfort zone of high biological productivity. Satisfaction, pleasure, assurance. In our not too distant past, open vistas of gleaming plains could give us clear sight lines on the whereabouts of game we could hunt, or the location of potential enemies, while the presence of clumps of trees, or forest edges, could ensure our shelter and concealment. Satisfaction, pleasure, assurance. Rivers lead to lakes or the sea edge, all good places to live, where food and water are plentiful. Satisfaction, pleasure, assurance.

"Perhaps, then, our sense of beauty in Nature originated from the effect on us of those places in the landscape that our species learned to understand best guaranteed the conditions for human wellbeing.

"More extreme examples? Well, even a dense jungle can give shelter and food and hiding places. Sometimes so can wetlands.

"High mountains suggest or provide refuge, convenient lookouts, hiding places for small communities that seek safety and survival in locations otherwise hostile for humans. If, in an inhospitable open place, we can't find a mountain, perhaps we can make one by building a castle and then, ages after the castle is deserted and ruined, and actually comes to resemble a small mountain rather than a building, we will wonder at it as a place of nostalgia and romantic beauty.

"Of course, in the sophisticated cultural world we have invented for ourselves, sounds have become poetry and music, colour and distance have become painting, tribal myths contain history and religion, as also do many novels. Of course, our arts have come to have lives of their own that are powerful and compulsive and complex – and will continue to progress far beyond their roots in ecosystems. And, of course, memory transformed by speech and writing has become the means whereby I can sketch for you this view of Edenish ecosystems. In all of this modern world, though, the roots are still there.

"Those roots must not be allowed to die.

"We may one day reach the stars. But if, finally, we turn our backs altogether on the Garden, on the ecosystems of this world, we will lose our roots, and then our hearts will shrivel. As a scientist I want to understand as much as I can about the ecosystem-Garden. As a person I need the continued existence of natural ecosystems – the kinds we evolved in and inhabited for thousands of years. The kinds we developed our binocular vision in, and our flexible hands and fingers and opposable thumbs that enable us to do complex tasks – like science and engineering and architecture and painting. The kinds that gave us the brain and its wiring plan that makes us what we are. I need those ecosystems and so do we all, whether we know it or not, even if our particular patch within the ecosystem of our culture seems to us to have been long separated from now irrelevant roots in behaviour and dreams, which began in times when we and Nature were more unified.

"I believe we will need them even more in the future as the heedless positivism that drives our 'value-free' technology continues to make inroads into the wondrous and still mysterious natural biological systems we all are heirs to."

We had reached what seemed like the endpoint, and some students were already moving towards the door, when someone said, "Dr Logan, could I ask a last question?"

I looked towards the sound of the voice which came from a tall dark bespectacled youth of spare build, standing somewhat apart from others against the left hand wall.

"Sure," I said

"The thing you haven't mentioned is ownership," he said. I listened hard, because I could guess what was coming.

"I mean, all the world is owned. It's either owned by individual people, or families, or it's owned by companies or industries, or it's public land – which means it's owned or controlled by the government. There's no unclaimed land anywhere any more. How do you propose to carry out programmes of ecosystem conservation if any owners can say to you that the land is theirs, to do with it what they want?"

I nodded. "Your question's right on the money. And the answer is not simple. If there's an answer possible it's along these lines: most people don't own land – except perhaps the little bit of ground they have a house or an apartment – or maybe a mud hut –

on. But whenever people own a substantial piece of the earth it can only be because it's been given to them – in most cases by their parents or other relatives – or they've purchased it from someone else. So, yes, that way, in the kind of world we live in, they 'own' it and . . . can more or less do what they like with it within pretty wide limits. So, as things stand at present, and have done for a long time, you have to try to inculcate a doctrine of 'wise stewardship.'"

"I don't get you," he said.

"I mean that, since there's no more land being made, it can never seem really fair that this, of all possible commodities, can end up in the hands – most of it, that is – of a minority of the population. Therefore, what we must argue for, and hope for, is that owners will come increasingly to understand the extremely special nature of their relationship with the earth, its uniqueness, its subtlety, and the responsibility that their ownership brings, along with its privileges, to understand that there will be special expectations that they will be good and enlightened stewards, and enable us all to benefit from that. If they don't do that, and if they never come to understand what the true nature of their bond with the earth ought to be, then, well, I think the future of all the things people like me are hoping for will be bleak."

The tall student, nodded. "Thank you."

The room fell silent. And then I said, "I often think that what the conservation movement needs is a spokesman of vastly greater eloquence than any of us at present seem to possess.

"I mean someone with a special kind of enlightenment and an inspiration that will make it shine. Something that will awaken people to this cause, and rouse them to spread the whole idea of conservation far and wide. Because, you know, this isn't just about learning, or science, or being well-informed about the world. It's about love of the earth. And it's a question of ethics."

There was a silence as the gathering broke up and we moved out of the room.

*   *   *

"You made quite a splash with your Gustafson lecture and the talk to students," said Rudy, back at Merriman a few days later. "It was hardly what they were expecting. They thought they were getting a straight-arrow scientist who would be Gung Ho! about the cutting edge of ecological research. Instead, as someone told me in a fairly fervid telephone call: 'This fellow gave us prophecy, metaphysics. Practically poetry. Pretty lyrical stuff. Took us all by surprise and at the end we nearly forgot to applaud. Most unusual goddamned Gustafson lecture we've ever had here as far as I'm concerned – and I've been attending them for fifteen years. Keep your eye on this guy, Rudy. He's certainly different."

I was not quite sure how to respond, but Rudy said, with a grin, "I think many of them probably thought you were really very subversive, but it seems to have gone over. Congratulations! Oh, and let me have a copy of the lecture, please."

Chapter 28

"My question, Dr Logan, does not relate to the – ah – in-invertebrate creatures that inhabit the bottoms of streams –" regardless of the very long and detailed statements concerning their biology by Dr Erlenmeyer and yourself, statements that are, in their way, no doubt authoritative and, to the expert, perhaps even interesting. I want instead to touch again on a very much more simple matter. Did, or did not, you and your colleagues and team of assistants collect water to enable an analyst to measure the levels of zinc in Dixon Creek and the Onyx River?"

I braced myself, felt I was frowning, knew my teeth were clenched because my jaws were aching. I tried to keep in mind the advice given me by our lawyer, Geraldine Howkins: "Appear calm and confident, not fretful and rattled. Remember, you are the expert."

"Of course we collected water and analysed it for zinc," I said. "I've answered that already. We – "

"Thank you, Dr Logan. A simple 'yes' will suffice."

My interrogator, who was called Williamson Montague (why, I wondered, was his first name not his last name?), now swung gracefully on his heels to face the judge.

"We have listened to extremely long statements, your Honour – for a day and a half, in fact, by two gentlemen of undoubted learning in their abstruse field of knowledge. These two are expert witnesses for the Town of Wystan. However, for those of us – a regrettably large number of us, I fear – who lack a formal education in the science of biology, their accounts are, quite simply, largely incomprehensible in the density of their explicatory material and the prolixity of their highly technical language. Those of us without their specialist knowledge are not equipped to follow such complex arguments as they have presented.

"This is a court of law, and here, if it is at all possible, we must deal in ideas and values that persons of ordinary education and intellect can grasp."

I looked at him in wonder. I was not used to seeing lawyers at work, and I had to admit his performance was admirable, if only for its smooth gall.

"It might be," he went on, "that if we were all as intelligent and informed as the learned Doctors Logan and Erlenmeyer, their sophisticated account about the tiny creatures that inhabit Dixon Creek – or I should say do *not* inhabit Dixon Creek – would appear, to every one of us, as convincing as evidence of damage to the Creek, as it apparently is to them. But their arguments are, I submit, just too needlessly complicated. For after all, everyone can follow the idea that water was collected from Dixon Creek, and that tests on that water produced clear, accurate figures that denote the levels of zinc in the water. And the tests show that only for very brief periods – less than a full day in duration – did zinc rise very high in Dixon Creek following flood rains. After the rains were over, the zinc levels fell right back down again to quite low levels. So it's very, very unlikely – and here I stress that I'm accepting, as a standard for comparison, well-documented, fully published, and generally acknowledged scientific standards – yes, it's very unlikely the zinc levels that were detected in the creek for less than one day could have killed fish."

The quiet firmness and self-confidence of Williamson Montague's delivery was all-important to his case. Geraldine Howkins, who had been retained by the town of Wystan, was a very able lawyer. But in those days, she was a "woman lawyer" up against a local legal icon – Montague – whom the folks enjoyed seeing whip the ass of eggheads like Erlenmeyer and me. Especially after he had no sooner dismissed our presentations as essentially unintelligible, but then proceeded to use "scientific" evidence of his own to disparage us.

It availed us nothing to point out in the most forceful way, as I did, and later Geraldine Howkins did, that we had definitively demonstrated there were no fish inhabiting the twenty-five miles of Dixon Creek below the mine site, and that, when placed in cages in the river just below the mine, fish died within a few days. And that even in the last pollution-zone station, that was twenty-five miles downstream from the mine, only half the exposed fish survived for two weeks.

Williamson Montague airily waved these facts away.

"So," he laughed. "You say, Dr Logan, sir, you say the fish died of zinc poisoning, even though you admit the level of zinc in the creek only three days after the flood could not, according to the published standards I have cited, possibly have killed the fish?" He laughed again.

"But look," I said, "in the creek there was always *some* zinc, and – "

"Thank you, Dr Logan. I have no further questions."

I had been trying to say that the published experimental results did not relate to exposures to lower concentrations of zinc for periods longer than one day. I wanted to say that our results could be explained by a few published studies that had shown even very low levels of zinc could kill fish in longer exposures.

But Williamson Montague dealt with us in masterful fashion, and the Town of Wystan lost its case against The Dixon Creek Mining Co.

Roly Templeton was consumed by fury against Monatgue and the mining company and, for a while, he was almost as mad with Erlenmeyer and me. We had failed him "after pretending we had such a hot case that failure was impossible."

Later on, he forgave us. The fat manager of the Town of Wystan was a decent and honest fellow. He knew we were not to blame. Montague had persuaded the court to believe that it was too ignorant and stupid to accept our arguments. He had also cleverly managed to paint me as a 'foreigner' – even perhaps a 'foreign mercenary', insincere, probably dishonest, parachuted in for the occasion, and also an inhabitant of decadent California. In sum, too slick and alien for the good, simple folk of Virginia. And he had also managed a subtle, sexist disparagement of Geraldine Howkins, who apart from being a highly intelligent lawyer was a slim, pretty and extremely young-looking woman (she was, in fact, thirty-two). Somehow, he was able to be a big tough lawyer man and also an old-fashioned, half-mocking flatterer, and a gentleman who never failed to address her with a cloying, old-world courtesy.

"We were raped by a master – all of us," said Myron afterwards, as he calmly lit a cigar and managed, somehow, an expression of rueful admiration as he watched Williamson Montague's tall, elegantly stooped figure drift through the court doorway.

I remember turning to Myron and saying that "In a criminal case this would have to be like finding not guilty a killer who had been caught in the act of shooting his victim through the heart. So why, for Christ's sake, did we lose?"

Myron sighed. "Yeah, well in our case there was apparently an inability to understand that science is a progressive enterprise, that because a standard value has been set down in printed form and followed for years, doesn't mean it can never eventually be superseded, that other ways of viewing a problem will arise which will be difficult at first for many non-scientists to accept. And ... there may have been a certain reluctance to penalise a profitable business venture like the mine that brought a lot of well-paying jobs to this region."

"Look, a child could see that this thing was about reluctance to pay for what – "

"You're missing the point, boy. To decide for us, the court would have needed to set a precedent. That would have looked like the court was pointing the finger not only at Dixon Creek Mining Co., but was making it possible to get at all sorts of similar operations that employ decent folks."

"So," I snarled in disgust, "how can we ever win?"

"Oh, we won, boy. Had 'em cold. Everyone knows that, Bill. Those local papers that aren't ultra-conservative rags have all been for us. And the radio. No, Bill, we won. But for legal justice we've just got to wait. The magic ingredient is time. Passage of time. In time, some judge with idealism about the law, and balls, will make the sort of tough realistic decision that has to be made. I mean on some other issue. After that, people like Dixon Creek Mining Co. will be on the run everywhere.

"You think we lost, which I understand. But we won. And in terms of the impact we made we've done better by losing."

"How the hell can you say that?"

"Simple, boy. Because everyone with half a brain can see we won but were just 'home town decisioned'. Our case, had we won it in court, would have had the air of a clean, neat, final judgement. But penalties would've been set at some amount that'd be trivial compared to the real restoration costs. The newspapers would have given an approving nod, the local conservationists would have patted themselves on the back, and things would have reverted to what they had been before, without a change in the law, or a generally perceived need for a change."

Myron slapped me on the back. "Because we lost something that by all rights we ought to have won, people will be upset. They will be angry. Dixon Creek will be famous. Its story will be across the country before long. I predict you and the IEC will receive much kudos from 'the Dixon Creek affair'. Oh, I dare say Roly Templeton, and Jack Schmidt, too – as head of the local conservationists – will feel very disappointed and humiliated and angry. But that's good, too, because they'll work like hell to get the laws toughened up.

"Don't fret, Bill. We won. We lost the battle, but the war will be ours. And on a grand scale, I believe."

For a while, Myron's assurance buoyed me so that I felt more optimistic about the entire future of conservation. He seemed so calm and logical and full of wisdom about how things would go.

But, of course, today's reality can always make a horse's ass of yesterday's assumptions – no matter how plausible they may have seemed at the time.

Chapter 29

Flight of Fancy was a box office hit, though not an over-whelming one. It did receive handsome reviews in many newspapers, including The New York Times and the Boston Globe. But then it was characterized as "an overblown fable" by The New Yorker, and there were other reviews that were similarly critical. But Betty's performance, as the obsessed and romantic aviatrix, was consistently, and sometimes lyrically, praised.

A month or two after the film's release I saw Irve Robichaud again in the Cosmo Studio canteen. I was finishing my third cup of coffee – which by now had grown cold, waiting there for Betty, who was having a discussion with the studio heads over a possible new contract (she declined it, to their chagrin).

I could not have foreseen that I would come to think much of Irve Robichaud, but underneath the slightly theatrical and very American stylization of his dress and demeanour, I had discovered, after several meetings, an honest and even empathic man. I don't believe he was truly interested in what my work was about, though he asked me a number of apparently serious questions. But certainly he liked Betty – considering her to be his discovery – and he seemed very pleased that our marriage was evidently holding up.

As we sat over more coffee – his first, my fourth, he blew

cigarette smoke in his sort of languorously elegant manner, and talked about *Flight of Fancy.*

"See, Bill, it isn't that the movie was that great. It's just that Betty is probably the best film actress in the business."

I must have stared at him because he waved a vague hand in my direction and said, "No, look, this isn't just me inanely mouthing studio publicity. You saw the goddam film. Don't you, for Chrissake, think she was tremendous?"

"Of course. Of course. I always did."

"Well, you were right, boy, right. It's bandied about by directors that she's the one-in-ten-year talent. She's a marvellous visual presence . . . though of course she's a commanding verbal actress too. Anyway, just about everyone in the industry – at least in the English-speaking world – thinks she's the most talented actress presently at work."

"Have you told her?"

"Told her? Jesus, of course. Everyone's telling her." His eyes narrowed and he glared at me. "Hasn't she told you what people are saying?"

"Something about it," I said. Which was a lie.

"Anyway," he went on, "she's wowed the critics, and" – he paused to blow more smoke, drink more coffee and smile thinly – "it's because of her the film's made money – and the studio bosses know it. They'd kill to keep her. Better watch out, son." He laughed. "But the irony is that the films she's in may never be real over-the-top smash hits."

"I don't get you," I said.

Irve looked at me curiously. "Yes, well I tend to forget that though you have a wife who will soon be one of the reigning queens of the silver screen, you yourself aren't a man of the movies. See, it's male actors who are nearly always the heavy hitters in selling films. The gals, no matter how good, how respected, haven't got the same box office heft. Ridiculous, but true. And the money they can ever make reflects that truth.

"Of course, I don't mean her films won't be sizeable successes. I'm positive they will be. And she'll make a fortune.

"Anyway, what d'you think of what they want her to do next – whether or not she signs a long-term contract with them? I don't just mean the part. I mean shooting on location in Bolivia?"

"Well, I guess it's okay. No. I mean I'm fearful about it.

Wouldn't you be? And three months there they say."

"I can say one thing, Bill, the preparation for this film will be serious and elaborate. I know the scriptwriters and they are going to do a good job. Men with a mission, really. They won an Academy Award for their work on *Dust and Gold* three years ago, and they're greedy for another. One of them – Brian Abruzzi – used to be a history professor at UCLA, so he's capable of sound scholarship in addition to being creative. And the basic story of Simon Bolivar is exciting enough as straight history. It could be a great movie."

"But Betty won't be –"

"The male lead! Right. But she will co-star. And it's the sort of thing that – with her looks and ability and those moves that only she can make – she'll be dynamite in. She's so . . . " He looked down quickly without finishing, and the thought dawned on me that he perhaps wasn't entirely comfortable saying these things that sounded too much like studio publicity. I was pleased I could detect as much delicacy in one so steeped in studio hype. And then, as I watched his downturned face and averted eyes, a second and more disturbing thought came to me about Irve Robichaud.

Chapter 30

Betty and I were at dinner at the Hausers'. It was interesting to see Rudy's reactions to Betty now that she had gained fame. I mean, here was an already acclaimed scientist, well-travelled, with world-wide contacts and colleagues, widely read, liberal, cultured, casual in everyday manners but high in the pantheon of scientific affairs, one who could count Nobel laureates as close friends. He was also married to a handsome and charming woman, herself a noted professor of social anthropology. Yet he was as beguiled by Betty – by her youthful beauty, striking physical presence, and above all that she was now that glittering American social icon, a film star – as any fame-besotted teenager.

He perhaps thought of me, also, as one to marvel at. What had I, in God's name, one of his several still youthful sidekicks, ever done, or was ever likely to do, to have attracted so unlikely a creature as this – no, more, to have persuaded her to marry me?

After dinner Nancy Hauser took the dishes to the kitchen and Betty went after her to help her clean up. They got talking and were obviously going to spend at least a few minutes together. Rudy, a glass of brandy in his hand, came up close to me, his face unaccustomedly flushed, his air more personal and confidential than I could recall.

"Bill, I have to tell you, that dazzling creature you've corralled into marriage, my God! well, none of my damned business, of course, but it does feel strange to meet her at one's dinner table, just like that. And find yourself talking everyday things with someone who's been this sort of 'glowing presence' in those movies. I guess, for you, she's just . . . well, the girl you married, but . . ."

I listened, half-listened, to his disproportionate admiration while I tried to plumb its meaning. I thought, "Jesus! He's not just this big, tough, legendary thinker of an ecologist after all. He's also still a teenager with a somewhat refined Jane Russell complex. Don't American males – of whatever intellectual calibre – ever grow up?" And then I remembered he was, after all, a Californian and inevitably a child of the Hollywood movie culture, like millions of his compatriots. And then I admitted to myself that I was saved from the discomfiture of having to compare my own enthralment by the smug feeling of knowing it had struck me when Betty was still an unknown fledgling stage actress.

When they came back from the kitchen, coffee cups in hand, Rudy turned to Betty.

"Well, Betty, from what you've said it seems you'll soon be busy with this new film. How long will you be in Bolivia?"

She hesitated and glanced at me for a long moment with a look that might have been wistful. "Four months, we now think. And then there'll be a lot of shooting on the studio lot. Maybe another six weeks. And I'll have to be on hand for the editing and dubbing for quite a time after that."

"So, the best part of a year in all."

"Something like that."

Rudy looked at her sideways. "He doesn't even know it yet, but I'm just about to ask Bill to go overseas . . . to Australia."

I looked up in surprise. "Nice of you to tell one of us, at least, Rudy."

He laughed. "We'll talk about it in a couple of days." Then he went on, "Well, I can't help worrying about the two of you. If you are going to be tied up with this film for most of a year, and Bill is away during the same period, you mightn't be able to see each other for a long time. Altogether too long, I'd say." He looked at me, then. "Just what about that, Bill? I don't want to pry, but I feel concern."

Neither of us seemed to know quite how to respond; after a

pause he said, "What I'm wondering about is whether you, Betty, could manage a month's holiday, when you could go to Bill while he's in Australia. I mean before this film in Bolivia gets under way. Do you both good, as you don't appear to have had much time together since you were married. Bill would have to work some, but I think you'd find plenty of time to be together."

"It would be great," Betty began.

"I want to avoid any appearance of seeming to meddle in your affairs," Rudy said, hurrying on self-consciously, "but I feel concern for Bill as a colleague and friend, and for you as his wife. Your jobs may put a lot of strain on you unless you're able to figure out some strategy to handle things."

"Well, we thank you for that, Rudy," I said, "and we couldn't agree more, of course." I glanced at Betty. "You probably could get away for a while like Rudy says."

"Maybe. We'll try to do something."

* * *

We had driven into the hills behind San Francisco and spent the afternoon rambling there. The day was blue and gold and clear. There was more than a hint of fall in the air, but it was a day of beauty and purity that made the world feel young, a place of pleasurable beginnings, and hope.

At five-thirty we stopped and lit a little campfire. I took food from a pack and we cooked a meal of lamb chops to eat with rolls, cheese and tomatoes – an Australian sort of picnic meal. We drank white wine and dark, strong coffee. The stars were soon out, silent and sovereign, far enough away from the city lights to blaze from the sky. We left it so late that we only just managed to find our parked car.

Betty took my hand. "Today was lovely, but otherwise things haven't been what we thought we had a right to expect, Bill."

"Oh, look, darling, because we got married we didn't suppose our other expectations would be set aside, surely?"

I thought I could just make out her hesitant smile in the starlight. "I just wonder if I can take this, Bill. I needed you very desperately during *Flight of Fancy*."

I tried to see her eyes, but it really was getting too dark to make them out properly. I said, "Look, Betty, when it comes to love and marriage I'm just a simple soul. I love you. I assume we've got a lifetime ahead of us. In a few years, say up to five, I should be finished with the sorts of things Rudy expects of me now. Whether I'll be with the IEC or some other operation, I'll want to settle down more. Maybe you, too, will find that all the travelling you're doing now will slow down, and most of your movies can be done hereabouts. For me, it's enough that I found you – I mean before someone else did. Cheer up. We can make it."

She gripped my hand and kissed me.

Deep down, of course, my confidence was far from being as solid as I was trying to make out.

Chapter 31

Two days later, Rudy came to my tiny office and parked himself in the only visitor's chair. "What do you know about marine turtles?" he asked.

"Well, I know there are several species and the biggest is the luth, or leathery, that can weigh up to half a ton, and I think the green turtle, which is much smaller, but still a big animal, is the most numerous."

"Right. Did you ever see green turtles before you left Australia?"

"No. They're on The Great Barrier Reef. But I never got there."

"Well, here's your chance. FAO and the Australian Government have asked IEC to undertake an ecological study of the green turtle – *Chelone mydas* – on The Great Barrier Reef and the coast of Queensland. You're the only Australian so far who's a member of IEC. The job's there for you if you want it."

"Sounds good," I said. "But what's it about? I mean, why do FAO and IEC want it done?"

"The green sea turtle's got a circumtropical distribution. In all the species of sea turtles the females come ashore to lay their eggs – and then they're easy to get at. People have killed them for meat and their eggs for maybe tens of thousands of years. The Barrier Reef turtles are probably pretty safe at present but, with the population explosion, their future in poor countries has long ago come under serious threat. If the trend worsens green turtles could face extinction in many places."

"So what is wanted? I mean in Oz."

"Initially, a population and behaviour study on a reef island off the north-eastern Australian mainland, where they lay their eggs. You know, estimates of eggs laid, numbers of beaching females, attempts to get age data, ideas of movements and so on. That's never been done before."

"And that would be considered to be a sort of baseline for comparison with other, exploited, areas, to give an idea of the impact of humans on reproductive success?"

Rudy nodded. "Only as an approximation, of course. What's down the road will be management of stocks to conserve them as well as exploit them . . . like fish stocks. But on the way we'll get a lot of fundamental ecological understanding of a wonderful animal."

"Okay, Rudy. I'm sold. When do I go?"

"Soon. We've already organized liaison with universities in Queensland and New South Wales. You could be a visiting research fellow with their zoology faculties."

"I have to ask why their own faculty don't do this work."

"Simplest possible reason, Bill. They can't seem to raise the research budget for it. There are several American groups working on Australian animals. You may be an Australian, but as far as Australia is concerned, you'll be coming to them with money from here."

I sighed . . . disgustedly.

He went on. "IEC and FAO will pay you a salary for the year it's reckoned it ought to take to launch this work. I want you to set the thing up, get it going properly, and find some good local people who can continue it after you leave. IEC would fund the work for three years if something satisfactory could be arranged. You'll have research funds sufficient for you to supervise a student or two if they'd permit that, and I expect they would. Students could help you, and get field experience themselves.

"Sounds good," I said.

Chapter 32

We were on a coral cay of the Capricorn Group of islands, at the southern edge of The Great Barrier Reef, and we waited in unmoving silence as the turtle struggled up the beach, which was formed entirely of coral fragments. She took an unconscionable time about it, but we had reason to be patient. Most turtles were easily disturbed by movements and would turn back towards the water. But this one was advancing almost directly towards where we sat on the soft warm sand in the gloom of a tropical evening. Eventually, she pulled up literally alongside us and began to dig her body pit. We never moved. Often after that, I was able to settle down beside turtles once they were fully engaged in digging, but this was the only time that, in selecting a place to dig, a turtle seemed to have no more regard for its audience than if we had been rocks or tree stumps.

My companion was a fellow biologist, Larry Marlowe from Sydney University, who was familiar with the reef. I had persuaded him to accompany me on this, my first visit to the domain of the turtles. The turtle was right against Larry and pretty soon he was being covered by coral sand flung out by her digging foreflippers. "I'm getting out of here," he muttered. "This bloody thing's going to bury me."

"Hold still," I whispered, "or I'll bury you myself. This is a chance in a million."

Actually, we were, both of us, amused and delighted. This would be a story to dine out on.

Twenty minutes of digging the body pit as the turtle gradually lowered herself below the surface of the beach, and then excavation of the egg chamber began. With dextrous, hand-like scoops of her hind flippers, she dug a compact, well-shaped chamber to the maximum depth allowed by the length of her flippers. On completing the egg chamber the turtle settled down to do what she had struggled ashore to do: lay eggs. She was oblivious of our movements, as if meditating on her own inner processes as they geared themselves for the release of about a hundred white, soft-shelled eggs, of the size and appearance of ping-pong balls.

Brushing off sand, we stood up to see it all more easily. We shone a flashlight below the turtle's short tail into the filling egg chamber, and looked into her teary eyes and at the blunt, armoured head. Nothing seemed capable of disturbing her now. We bent to examine several deep, bleeding scratches on the soft underside of her flippers, caused by the jagged coral and beach rock over which she had dragged herself as she lumbered towards the sandy beach.

Beside us, ignoring us, intent on the climactic task of its life, was this large creature that, according to the fossil record, hardly differed from its ancestors of two hundred million years ago; a creature more constant in form than the beach rocks it had struggled over. And here were we humans, members of a species of very short ancestry as vertebrates go, a species self-anointed as creation's peak, alone on a beach at midnight, thinking and talking about turtles, trying our best as professional biologists to comprehend all aspects of our relationships with turtles as organisms.

And next day we continued to discuss green sea turtles, whose evolutionary lineages were a hundred times longer than our own. And gradually we found ourselves talking about matters of killing and death. The green turtles don't kill members of their own species, or other vertebrates, and the larger turtles are safe from nearly all marine predators, except, sometimes, very large sharks. But many predators attack the juveniles which, on hatching, after an incubation period lasting from six to twelve weeks, depending on sand temperatures, push their way up through the sand-plugged egg chamber and through the additional couple of feet of sand that their mother dragged back to fill in the body pit. At last, the hatchling turtles break out all at once, almost explosively, and waddle furiously down the beach towards the sea, like a horde of the cutest imaginable mechanical toys. Sea birds kill some on the way, so do sand crabs. In some parts of their world range rats attack them; also feral dogs and pigs, which will excavate the egg chambers and eat the eggs or the hatchlings. The young turtles, when they first hit the sea, are buoyant, unable to submerge, helpless targets for seabirds and predatory fish.

But the greatest threats to green turtles are human beings who have long killed turtles and eaten them and their eggs. Humans kill female turtles when they are ashore, and even the males which never come ashore. Often, on capture, these large beasts are transported on their backs, alive, for long distances, then unceremoniously slaughtered for their meat. Victorian Englishmen relished green turtle soup, salivating over it, washing it down with gulps of claret, supping on the essence of a creature that might have lived another twenty years, mated a score of times, produced thousands of eggs and cruised, cruised endlessly, as if in rapt and dreaming contemplation of the clean clear waters of green and blue tropic seas.

* * *

Thinking of the fates of the turtles made us reflect on the ways and means of dying. And we found ourselves asking whether it could have helped humans in our attitudes to each other if, like the turtles, we had been of a more robust construction, our bodies well-protected against all but major impacts. Would we then ever have bothered to strike at one another with sticks and stones? Would inventing weapons ever have occurred to us? For though we are pathetically fragile creatures, killers lurk within us. We are these delicate sacs of protein and water, suspended on skeletons that a dog's jaws can crack. Perhaps this is why the demon in us tempts us – on the slightest of pretexts – to crush one another like blobs of mucus.

Perhaps the turtles will survive long after we feeble, watery and violent beings have departed. The "wisdom" of the turtles lies in their having evolved a body form that, in contrast to our own, seems closer to the lithosphere than to the biosphere . . . more like a rock than a propped up jellyfish.

Chapter 33

I did not return to Sydney with Larry Marlowe but went instead to visit the University of Queensland. On my second day there, over morning coffee, I was talking with Richard Petrucci, one of the zoology faculty. I had explained what I hoped to be doing with the turtles and he said, "Well, but do you mean you'll be working with Andrew Peacock?"

I confessed ignorance of Peacock's identity.

"I see," said Petrucci, surprised. "Well, no doubt you'll be meeting him before long. I mean, it's inevitable. Peacock's a post-doctoral research fellow at North Queensland University. He's been working on green turtles for a year and a half. I think what you're proposing to do must be just about the same as his project. I hear he's an able chap. I'm pretty surprised if no one has told you about him."

I think – I am sure – that my mouth was gaping. Those near us who had heard our conversation were smiling nervously. One or two muttered something. "I have to tell you," I said at last, "that till this moment I'd never heard of Andrew Peacock. Can you tell me how to contact him?"

* * *

I flew north a week later to meet Andrew Peacock on the coral cay on which he was based. It was fairly embarrassing. A Ph.D. from Oxbridge, he had come from Britain on a lucrative and prestigious fellowship, specifically to study the turtles. He had come to North Queensland, and quickly devised and mounted a programme that was working well and was eerily similar to the kind of thing Rudy Hauser and I had envisaged.

How did it come about that the Australian government had failed to advise us of this?

It took me several days of laborious enquiries to get a simple answer to this question, which was that there had been no obvious way for the connection to be made. Peacock had been accepted as an already-funded researcher with an excellent academic record. Although he had been at work for a year and a half and was expected to publish several papers on his research eventually, nothing had appeared in the scientific journals at the time we met. North Queensland was a remote location and few zoologists elsewhere in Australia were aware of Peacock's presence.

All that was about to change.

I found Peacock an intelligent, highly motivated, self-propelled Yorkeshireman. He was also arrogant, vain, and critical of others. Personally, I got along fine with him – probably because he sensed I was totally unimpressed by his bullshit, and perhaps also because he soon realised I had no intention of competing with him. He had already shown great energy and enterprise in building up an extensive network of people who would help him in his work, and had made all the preliminary arrangements needed for the logistics of his project. He knew of the IEC and was respectful of me as one of its agents. But I could see and appreciate that he was an aggressive and independent investigator, who believed he ought to be left alone to get on with what he had long planned and was already deeply involved in.

I agreed with him, but it was not easy for me. My single visit to the Reef had left me fascinated by the turtles. It now appeared unlikely that I would ever have an opportunity to study animals of such immediate allure: great docile beasts, at once eminently approachable and essentially mysterious. Yet I knew that Andrew Peacock would do at least as good a job as I could have, that he had passion and intensity and imagination, that he was an expert field worker, and totally alive to all the problems and challenges before him.

*   *   *

I made a long, expensive phone call to Rudy a few days after I left Peacock. I told him everything. "It's like this," I said, "for reasons that are really nobody's fault, we've got a guy here who's going to do a bang up job on this problem, and do it just about like we planned. I could hang about here and make a nuisance of myself, but I don't think IEC wants me to play second fiddle to, or compete with, a competent scientist on a project he's already made his own. That would just be wasting our resources. Don't get me wrong, Rudy. I'm nuts about these creatures, and it would have been a sweet project. But I think we ought to chalk it up as a no go, and just move on. I hope you agree."

Rudy, after a long moment of reflection, did agree. "Okay, Bill, drop it. Instead, I'm going to ask you to go straight on to Indonesia, where they want some critical evaluation of the status and future of 'tambak' fish culture operations."

He spent the next half hour explaining and ended with: "All arrangements for you to meet the people you'll need to see can be made from here before you leave Australia. But before that, maybe go see Peacock for a little while, just to give him a bit of field help with his turtles – if he'd like that. And then wish him well. Just a sort of friendly gesture IEC should make whenever possible, especially when it's towards someone who's doing the kind of thing we ought to be doing ourselves.

"Oh, and by the way, why don't you call Betty right away and ask her to have a couple of weeks holiday with you in Australia before you go on to the next assignment? She hasn't begun her filming in Bolivia yet, has she?"

Chapter 34

I met Betty at Sydney Airport and we embraced hungrily. It had only been three weeks since we parted but she had lost weight, was more slender than ever, and her face, beautiful as always, looked almost drawn.

I felt anxious. "Are you okay, love?"

She smiled. "Oh, you noticed I've lost a pound or two."

"Well, I . . . yes."

"It's nothing. This Bolivar thing calls for a lean and hungry looking female lead. I've – "

"You've been missing lunch and breakfasting on black coffee?"

"No! Nothing so stupid. Just eating a slice less bread and more cottage cheese and fruit. Anyway, it's over now. I'm as skinny as I need to get."

"Amen to that," I said.

I wanted to show her Sydney, and I did. She loved it, revelled in the space and light and ocean. Most Americans immediately compared it to Los Angeles. If anything, it was San Francisco it resembled; but Betty was quick to spot the hotter climate, the vast skies, the absence of sea fog in the mornings.

In our week in Sydney I showed her where I'd grown up,

she met my parents, we cruised on the Harbour day and night, dined near the water and surfed. A keen surfer as a girl, she loved the great Sydney beaches, the life-saving carnivals.

"My God, Bill," she laughed, as we surfed for the fourth day in a row, "I've just realised that in California the sun falls into the sea in the evening. Here, it disappears into the countryside. What the hell – ?"

"Jesus," I said, "all you Yanks are the same. Utterly ignorant of geography outside the States . . . if not there as well. It happens, my precious, that we are now on the west side of the Pacific, while California is on the east side. So . . ."

"Oh," she said, seemingly satisfied, though from her faint air of perplexity, I still could not be sure she got the point.

For two more weeks we travelled. We went to western New South Wales to see the great sheep plains and the edge of the desert; we saw red kangaroos, emus, white cockatoos and several hundred thousand of the country's one hundred and fifty million sheep. We spent two days in glorious weather tramping in the Blue Mountains west of Sydney. Then we went north to let her see the Barrier Reef and get a sense of what I had expected I would be doing.

We visited Andrew Peacock at his base of operations on the coral cay at the southern extremity of the reef – a fifty mile launch trip during which Betty demonstrated that she was a poor sailor, while I hung over the side watching two dolphins ride the bow wave of our launch.

Peacock, free now of any suspicions that I might be a competitor in his work, was hospitable and friendly, and clearly bucked to meet a young woman of extraordinary beauty and personality. He had even seen her films, and showed all the usual awe of an inveterate moviegoer in the immediate presence of a screen goddess. But Betty put Andrew Peacock at ease by being entranced by the turtles, and by showing great interest in his research (more than in mine, I told myself ruefully). Peacock's aggressiveness melted away, and I began to realise it was not a constant of his personality, but only a hedge against those he was unsure of. I began to like him.

Our third day on the island behind us, we sat on the beach after midnight with Andrew, Betty operating a pair of tagging pliers as she attached a monel metal tag to the left front flipper of a large female green turtle that had just finished egg-laying. She patted the turtle on the head and spoke to it in the British accent she had learned for ***Tono-Bungay***. "Off you go, old thing. Have a good life and a long one." We watched as the turtle lumbered back to the water.

"Bill," said Andrew, "leave Betty here. You can piss off somewhere else yourself. But I need a decent assistant about her age. And must be a woman; women are better at this sort of thing than guys. Oh, and must have raving good looks. Betty's a bit short in that department, of course, but she's not a bad kid overall, and I can make allowances."

Betty chuckled and he turned to her. "Listen love, stay here with me and we'll do a movie about turtles. It'll make box office history. We'll call it ***The Born Free Turtle.*** How about that? Oh, you can star in it, of course. But stay here with me. You don't want to go off to bloody Bolivia to make some load of codswallop about bloody old Bolivar. And I'll keep you company here so you don't miss Bill, who's going off like an idiot to bloody Indonesia, or somewhere."

It was interesting. She was a woman men liked. First of all they were attracted by her appearance, but after that they found they liked *her*. . . liked just the person she was. She had the face and form to attract men, but she was never a temptress. That, I knew, was what had really drawn me: that she did not presume on her remarkable attractiveness to be liked by men. I knew that even if she had been quite plain, her charm would still have drawn men. It was the way she was.

* * *

The Bolivar film loomed. She had to go back.

When she was about to fly from Sydney, she took my hand as we stood near the departure lounge and said, "It was lovely, Bill. A miracle. The kind of things we'd never really had. I'll never forget it. Oz is wonderful. So are you. But I don't know how much of this other stuff I can take. D'you think, maybe, we're 'star-crossed lovers'?"

"Come on, Betty, cut it out."

"Yes, but when we're together we're so okay."

"I take you to mean by that, that one or both of us shouldn't be ambitious about our 'career'?"

"Something like that, I guess."

"Okay, my dear. Which of us will it be? Do you really think

I'll try to prevail on you to give up films?"

"No, but – "

"Or that I'm likely to just put away my tools and become a 'husband in attendance', as wifey flutters around the globe in a blaze of publicity amid a dozen cigar-chomping film magnates and an adoring bunch of fans?"

"Bill, please," she was serious, "what is going to happen to us?"

I reflected. "I suppose," I said, "you could return to the stage, which would require far less moving about for you, and I could try to get a job in some place where we could establish a home. Let's think about it the rest of the year. You get this Bolivar/Bolivia caper over, and I'll go to Indonesia, and then we'll have another holiday – a longer one I hope."

Before either of us could burst into tears I kissed her and waved her goodbye as she entered the departure lounge. Twenty minutes later her plane had taken off and away she went into her other life.

I watched the plane till it disappeared from sight and sound, carrying my love far away from me. Very far.

My heart was a stone.

Chapter 35

Ten taxi drivers had sprung at me as I debarked from the airport bus. I picked the one who looked youngest and brightest; his companion, another young man, energetically seized my bags and, with a great flourish and a reassuring laugh, stowed them in the taxi trunk, the lid of which refused to stay closed and had to be tied down with a piece of hairy rope. The rattletrap taxi trailed a dense exhaust cloud as it moved forwards with a lurch, a creak of springs, and a roar that betrayed the absence of an exhaust muffler.

As we ran though the centre of Djakarta I saw it was replete with large celebratory open spaces that displayed gargantuan, stylized statues of heroic figures breaking chains of bondage. The spaces were flanked by huge, western-style, luxury hotels.

Within a mile or so the taxi ran up against the city's perimeter ring that seemed almost like a solid object so dense was it, and so sudden and complete was the transition.

For a few minutes I felt an uneasiness shading to fear as we edged out into the substance of the perimeter ring, whose apparent seething confusion was outside my experience. As we entered it, I felt as if I were entering a malignant forest, or even a monstrous crown of thorns. I was further disconcerted by the driver's baggage handler, who had sprung into the front seat at the last second and was now attempting to engage me in a lively conversation in clear, if eccentric, English. For those few minutes I felt a sense of utter alienation.

The taxi slowly twisted its way through the ring, often stopping for long seconds, sometimes for a minute or more, such was the press of people.

The ring was made of noise, adults, children, noise, trishaws, antique and decaying cars and trucks, crumbling buildings, dust, live hens, vegetable stands, noise, the pungent odour of mutton satay grilling over charcoal, shouting, bustle . . . noise. These were things I noted but could make little sense of in this, my first experience of a vast, overcrowded city of an over-populated country.

The stupefying heat demanded that the taxi's windows be open and people peered in, gesturing; children extended their arms into the taxi, begging. I imagined the vehicle breaking down, my driver and his friend going for help, abandoning me in the middle of this mad maze, surrounded by a million people, few of whom spoke English.

The mind is a monkey. Soon the driver, his friend and I were chatting amiably. I anticipated that the tip I would offer them would be inflated merely on account of the added gaiety of the friend's presence. I did not begrudge them; I knew I would feel relief in just getting to my destination without mishap, and in good company.

After half an hour the taxi had crept through the perimeter ring and emerged into the surrounding countryside. Now we forged steadily towards the town of Toorang, our progress impeded only by the slowness of many old and dilapidated vehicles, at least some of which dated from before the time the Dutch had left. We passed mile after mile of stopped vehicles of every size and shape. Human legs stuck our from under them as men toiled to get them running. There were fresh oil stains all over the roadside; there were wheels, axles, springs, sometime entire engines lying alongside the legs. Such scenes, as I soon came to understand, were part of the legacy of the withdrawal of the Dutch who, on quitting Indonesia, had left the country with a virtually destroyed infrastructure, with little or no technical and manufacturing sector.

Toorang, a half-hour drive from the perimeter, was hardly a glamorous place, but it had a large, rather pleasant, country town feel about it, seemed relatively clean and uncluttered, was not far from hills and mountains. It was the site of a college devoted to agriculture and fisheries, and it was part of my assignment to establish contact with the staff of the fisheries school. I was expected to discuss current fish culture practices and problems with the faculty members, and respond, if possible, to an FAO and IEC initiative that had followed an official request by the Indonesian government for an independent foreign expert to review these practices. I would also be expected to write a detailed report advising on ways in which fish culture in Indonesia could be made more productive.

I hoped, and looked forward to, further clarification of my task after meeting members of the fishery faculty at the college.

Chapter 36

The morning after my arrival in Toorang I met Dr Natsir, a handsome man of about forty-five, who welcomed me warmly. He explained in excellent, scarcely accented English that he was the immediate past-director of the fisheries school of the university and that, in half an hour, he would be introducing me to his successor, Dr Hakim.

"Dr Hakim is at present attending a meeting. He has asked me to convey his apology for not being on hand to welcome you." He gave a short laugh. "Dr Hakim is a young man – only thirty-three – but since he became director his duties have greatly increased. He says, now, that he feels much older than I look." This thought seemed to please Dr Natsir, who laughed again.

"I'm sure we'll all be waiting keenly," he said, " for the report you'll eventually write about how we are doing things and why we are not doing them better . . . for FAO, is it, and for – ah – IEC?"

"Yes. For IEC – the Institute for Ecology and Conservation."

"I must tell you," said Dr Natsir, "that directions and requests concerning your visit came to us from on high. From a senior minister of the government. Maybe you will find Dr Hakim a little hesitant when you meet him. He does not quite understand what it is you will be expecting from him. And, though this is nothing against him" – at this point Dr Natsir smiled – "he has not yet had quite enough experience to know how to deal with matters such as this."

My impression of Dr Natsir as a self-assured and experienced gentleman was reinforced. *He* knew very well how to deal.

I decided it was time to change the subject which concerned people of whom I knew nothing and was taking me nowhere. "If you won't be embarrassed by a compliment," I said, "your English is excellent. So many non-native speakers of English make us feel fools over our own incompetence with other languages."

Dr Natsir nodded, smiling. "Any facility I have was acquired while I was doing my Ph.D. in the States." He continued to smile showing remarkable white teeth.

"And where did you study?"

"At the University of Georgia. Do you know it?"

"Know of it. I haven't been there."

"I studied there under Professor Ernest Williamson. Perhaps you have met him?"

"Only know of him. But of course he's a famous ecologist."

"It was a great honour to work under his direction," said Dr Natsir.

"And Dr Hakim, did he also do his Ph.D. work overseas?"

"Dr Hakim was at Virginia Polytechnic."

"Another good place," I said.

"And what about you?" he asked.

"Where did I study? Well, Sydney, Australia, for my bachelor degree, then Merriman under Rudy Hauser for the Ph.D."

"Professor Hauser is also a great ecologist." Dr Natsir gave an emphatic nod.

"Yes, he is, though about ten years Williamson's junior."

"So it is Dr Hauser who has sent you here?"

"Yes, in his capacity as director and founder of IEC, but at the request of FAO."

"Ah," he said, nodding as if a great weight had been lifted from his mind. "Then with such a distinguished background I am sure you will be the right man for this job."

"I wish I was more sure what this job really is supposed to be," I said. "As for the distinguished background, well. . . it's no more distinguished than your own."

I sensed at once I had said the wrong thing. His expression told me he was hoping I would be a person to whom he and his colleagues could, without hesitation, look for authority and wisdom.

He swallowed and did his best. "Now, now, Dr Logan. Your work on thermal springs is very well regarded, and also your studies on coyotes and on stream pollution."

I was startled at his knowledge and must have shown it.

He chuckled. "Oh, yes," he said, "we know of you here. Indonesia is not so remote as all that."

It occurred to me that Rudy had sent a personal letter to the fisheries school by way of an introduction for me. That letter would be the source of Dr Natsir's information. I began to feel a bit uncomfortable in his presence, as I waited for the appearance of Dr Hakim.

But Dr Natsir wasn't finished. He suddenly changed his tack.

"You are accustomed to high salaries, you Americans – and Australians," he declared. "It is different for us. I make the equivalent of about five hundred American dollars a year. On that I cannot afford a car."

I nodded cautiously.

"But," he went on, "our cost of living is low. Food is cheap. I and my family live in a fairly spacious house from Dutch colonial times for which we pay a low rent. My wife possesses a degree, but does not work outside the home. She could not do so if she wanted to. We have five young children."

Unsure of how to respond, I said, "I see."

It did not seem to matter. Dr Natsir was simply giving an address to a small captive audience. He had probably given it many times before.

"Married women in Indonesia stay in the home and raise families," he smiled. "It is a bit different in America, and I think, in Australia."

"True," I said.

"Here, women accept this."

"Well," I said, feeling unsure of myself, "until pretty recently most of them had to accept it in America and Australia. But I think that time may be over forever."

"Not here. Here, we do things our way."

"Well, I don't know about that, of course, but it wouldn't surprise me if the way you do things will change here too."

Dr Natsir shook his head. "It cannot," he said.

Our conversation appeared to have petered out, but then

Dr Natsir's tone and manner – which had started out as slightly arch, and then had begun to sound rather carping – changed abruptly, to become more business-like and direct. Suddenly he sounded like someone who could be a director of a scientific enterprise. "So," he said, "we all know, of course, that it is our tambaks you are being asked to report on."

"That's right," I said, "and I'd very much like to hear as much about them as possible, from anyone who can inform me. And of course I'll need to visit some tambak sites." Saying this, I tried to sound and look as appropriately humble and open as possible.

Dr Natsir gave a short, understanding nod. "Of course. And, well, while you are waiting for Dr Hakim, I can give you a few facts about them."

"If you'd be kind enough," I said.

"Well, let me see. To begin at the beginning," – Dr Natsir smiled; I could see he was enjoying his mastery of the English language – " tambak fish culture is based on capture of a marine fish, the milkfish – *Chanos chanos* – which is then reared to market size in essentially brackish-water ponds, or as we call them, tambaks."

He paused, almost imperceptibly drew himself more tensely erect, and looked at me in a slightly challenging way. "You may not quite realise that tambak fish culture – of which Indonesians are regarded as the inventors – came into being a couple of centuries ago. And though Westerners may tend to view them as the result of a kind of trial and error practice that peasants came up with almost by accident, the methods actually required sharp judgement and much ingenuity."

"I'm sure that's true," I said. What I did not say was that I had recently read and absorbed a comprehensive report on tambak fish culture in all its aspects, written by W.H. Schuster, a Dutch biologist. I did not mention the report. I wanted to see whether the account Dr Natsir was going to give me would go much beyond Schuster's, which was by now more than twenty years old.

Also, I thought that by mentioning Schuster at this point, I might awaken delicate sensibilities about former Dutch rule of Indonesia, which could cloud future communication. Well, I thought to myself, this guy, if anybody, ought to know the tambak scene. He's an Indonesian, he's got a U.S. Ph.D., and he's a local. Shut up and listen to him.

"The greatest part of the invention," Dr Natsir went on, "was that people saw that ponds could be excavated in the soil of mangrove swamps. That meant that the mud removed could be used to form the containment banks of the ponds, and because they were in a tidal zone they could be filled or emptied as required by the use of simple gates or retaining walls."

He paused slightly again, and looked at me in a way I could not quite decipher. I decided that praise would not be amiss. I said, "Yes, that was a great idea, using a natural system like that. It –"

Dr Natsir gave a perfunctory nod. "The other good idea," he said, "was to use the milkfish. It is a marine fish of wide distribution in tropical seas, yet its young are easily captured in shallow, beach areas. And because tambaks are in mangrove regions they are not far from the sea, so it was always easy to transport the young to the tambaks. And then, the fact that the milkfish is tolerant of wide salinity changes is important, because tambak salinity levels are variable."

I nodded again, and murmured further praise, but Dr Natsir paid no attention. He was in full oratorical flight. He would probably have continued even if I had turned my back and walked away.

"Once the milkfish have been released in the tambaks to grow to market size, there are two types of plant food available to them, depending upon the kind of bottom mud in the tambak. One sort of mud is soft, biologically active, containing much organic matter. In a tambak of this kind, there will grow a carpet of soft blue-green algae, high in nitrogen – very nutritious for the rapid growth of young milkfish. In a second type of tambak, the bottom mud is firmer, less organic, encouraging the abundant growth of strong and wiry filaments of green algae. Milkfish can eat and grow on either type of alga. But the blue-greens grow continuously, ensuring a constant food supply. The greens, though they can be plentiful, are too tough for young milkfish to bite off. Only after they begin to decay and break up can the small fish eat them. Of course, because the fish can be removed alive from a tambak, it is sometimes possible, after their initial growth on blue-greens, to transfer them to a tambak in which they will now be large enough to crop the tougher green algae."

Dr Natsir's account had become a formal peroration, and I had remained silent for many minutes. I noted, though, that it parallelled precisely what I had already read in Schuster's account. This was, perhaps, encouraging. Now he paused. His eyes, which had appeared rather unfocussed, came to rest again on me.

I pulled myself together, smiled, and said, "Yes, that really is a fascinating story. I'm really looking forward to seeing tambaks."

I waited. I presumed I would now be told about significant recent changes in tambak management that had occurred since Schuster's report of a generation earlier. Nothing came from Dr Natsir. The silence very soon became uncomfortable. I decided to risk some questions.

"So, tell me," I began, "what has happened – ?"

Dr Natsir looked at his watch. "Well," he said, cheerfully, "I see it is time to take you along to Dr Hakim."

* * *

Dr Hakim, the director, was a plump, plain, serious looking man who wore large round glasses. He was pleasant but shy, with none of his predecessor's poise and manner. He looked a lot younger than his alleged thirty-three years. I felt much older than he.

He immediately asked me to address the fishery school's students. I asked when.

"Could you, perhaps, talk to them in an hour, say at eleven?"

"I have no slides ready, or lecture notes. I could do it tomorrow easily enough. Would that do?"

"Yes . . , " he said. "But today would be better. There is an excursion planned for the students for tomorrow. They are going to visit some fish ponds. A bus has been ordered. Today you could simply talk about international trends in fisheries, or something about fisheries in North America or . . . or just about anything."

I felt uncertain. I wondered what the students would be like. I might be covering ground already familiar to them. "Maybe I could talk about how recent work on population dynamics is affecting fisheries management in the Atlantic," I suggested.

"Excellent. Very fine. That would do nicely." His round, serious face almost managed a smile.

* * *

The lecture was held in the library – a large, high-ceilinged room in a building from the days of the Dutch. The shelves of the library could have held ten thousand volumes, but most of them were empty. There might have been five hundred books in all, scattered around the shelves in small irregular piles And when I browsed among them for a few minutes before the lecture, gulping down a cup of coffee, I saw they were mostly donated handbooks and pamphlets from FAO or other organizations, or else were obsolete textbooks.

The students filed in, dressed in light field clothes, having just returned from an excursion to a nearby river where, I was told, they had been collecting and tagging fish.

Dr Hakim introduced me, courteously but briefly, then excused himself: something required his attention, he said. The students seemed nice, clean, orderly, friendly, very respectful, attentive. They looked very young. Nearly half of them were girls; I wondered what Dr Natsir would think of that.

Halfway through my talk, which was proceeding in a silence and calm that were unsettling because of their completeness, I grew aware that a small group of persons had appeared outside the library and were crowding around an open window to listen. I assumed they were more students who had come late. "Come right in," I called. As I said it I realised they looked older than the audience in the library. I wondered if they were graduate students.

"It is Dr Hakim and other faculty members," said a student in the front row. Startled, I peered out. There stood Dr Hakim, with other young-looking people – and some less young, whom I now realised were his colleagues. Mystified and embarrassed, I went to the window. As I approached, Dr Hakim and the others drew back.

"Was there something you wanted to tell me, Dr Hakim?" I asked.

"No," said Dr Hakim in a faint voice. "We did not wish to interrupt. I brought my colleagues to hear you. Please . . . continue. We will listen from where we are."

Baffled, but not wanting to disturb whatever sensibilities might be in play, I returned to my talk. Some students were smiling as if to cover embarrassment.

At the end, a student representative thanked me profusely in fluent English, presented me with a small memorial medallion, and led the applause. The students filed out quietly, and the faculty members moved off, with the exception of Dr Hakim who, with some difficulty, began to climb through the window.

Nobody had asked any questions.

Chapter 37

 I was put up in a college hostel. The accommodation was plain, but clean and comfortable. The food was good and included an ample supply of fresh fruit. I met several visiting agricultural consultants from Britain and America, and evenings at the hostel were quite convivial, if short on alcohol – which was prohibited by law.

 Two young lecturers from the University – Mr Hamzah and Mr Saleh – were my guides. In the first day we covered many miles of country that, centuries ago, when there were many fewer people, must have been beautiful with forested hills and mountains and many rivers. But now it was a landscape shaped by the intense activity of a large rural population. An intricate terracing of rice paddy climbed high up the sides of the hills, transforming the countryside into something resembling a giant coloured relief map, with the terraces like contour lines. A sort of agricultural miracle had been achieved here, over the aeons, that by now assumed a species of that abstract topographic elegance common to many highly modified landscapes of Far Eastern and Asian countries, as endlessly celebrated in magazine photoessays.

 We paused at what my guides referred to as a fishery research station, but which was really just roofed over holding tanks for fishes. It was impossible to ascertain the intended fate of the fish that were large and beautiful carp of several species and seemed to have the status of exhibits in a zoo rather than experimental fish.

As for me, apart from lunching with Mr Hamzah and Mr Saleh at a pleasant hillside restaurant, where fountains played in shaded rock-gardens, and where we lingered long over our food, I was continually asking when we might see tambaks. Tambaks, as I kept politely explaining, were what I was in Indonesia about, what the IEC had been requested to pass a sort of ecological and economic judgement on, to devise new methods for, etc., etc., etc.

Just as repeatedly, even more politely, my guides attempted to reassure me. Soon enough we would be among the tambaks. But first, the itinerary was to show me more field stations, more visually splendid carp, visit a typical village among the rice terraces, etc., etc., etc.

Well, I had no objection. It could all be deemed information out of which I might eventually be able to conjure some kind of pattern. In fact it was fascinating to arrive at the "typical village", walk across the stamped earth of the village square, be introduced to an elderly, and wise-appearing dignitary (the mayor?) and be welcomed into his simple, but spotlessly clean, house. I shook his delicate, slim hand, while he nodded his small, neat head, smiled austerely and said some quiet words that I took to be a welcome. Around us stood village youths, gaping, but polite and friendly. Once again, I was shown some carp of formidable size and apparent health, this time in a large bucket. My guides spoke to people in their native tongue, evidently accounting for my presence.

I enjoyed the visit and felt these were probably an amiable people, while remembering having been impressed by the many armed police and soldiers I had seen on my transit of Djkarta and on the road to Toorang.

We began on the second day much as we had proceeded during the first day, but then, after two more field stations, late in the morning we were on our way to the tambaks, back towards Djkarta, towards the coast. At last . . .

Then our car ran out of oil. Not fuel. Oil. Engine oil.

Dr Hakim had explained that the college had no vehicles of its own. All of them had to be rented. He produced the car of the day with something like a flourish. This vehicle had been especially hired to convey me around the countryside and, eventually, to a tambak region, as a "distinguished scientific visitor" on my important mission. I appreciated the compliment and was impressed, not least because the vehicle was a well-shined BMW sedan in the charge of a peaked-capped driver in a dark uniform, who had delivered the vehicle and was to be our driver. I had never before been conveyed in a chauffeur-driven car.

I felt, however, a certain lowering of confidence as I sat down on the back seat of this fine-looking car and felt my arse crunch agonizingly against steel that was covered only by thin leather and collapsed springs.

I noticed, soon after we were under way, that gear shifting appeared extremely difficult for the chauffeur – and there was visible play in the car's steering.

The first bit of rough road told me that the shock absorbers were not working.

A bit later it became clear that the driver was unfamiliar with the vehicle and with the roads we were traversing.

A large, mechanically sound BMW would have effortlessly devoured the hills. Ours struggled up them in low gear like a fat man awaiting bypass surgery. Nearing the top of one steep hill, the car suddenly slipped out of gear. Only swift work by our driver prevented a disastrous run backwards. The brakes worked on that occasion, but I had already noticed that stopping was often something of a problem, judging from the bleak expression and tensed body of the chauffeur as we approached intersections . . . and the loud whining noise from the brakes every time we drew to a halt. It had already occurred to me that running downhill might eventually become a fairly interesting experience.

Then, at last, halfway up a long hill, smoke suddenly poured out from under the engine hood. The car stopped and we stumbled out. The chauffeur approached the engine. I was concerned for him. "I think you ought to be very careful," I said, jumping out of the back seat. "There could be a fire."

He knew a little English. "I must . . . ah . . . attend to it. I – I am responsible." He looked mortified.

"Never mind about that, my friend," I said. "Take care. You're a lot more important than this car. Let's just wait a few minutes till the engine cools off a bit." We waited for many minutes, until most of the smoke had abated, Mr Hamzah and Mr Saleh muttering softly to each other in Indonesian.

At last, the chauffeur and I lifted the engine hood. There was still a little smoke and the engine was very hot. I saw a trace of oil on the road.

"I reckon," I said, "most of the oil has leaked out of the engine." I looked at the dipstick; it was dry. I held it out to the chauffeur. "See that?"

He nodded gloomily. If I had been in charge of a car and something like this had happened to me, I would have been exceedingly angry with whoever had supplied the vehicle. Our driver had the look of someone who expected to be blamed. So I felt angry anyway ... for him; sorry for this poor man, dressed up in his chauffeur's outfit, given charge of a car in terrible condition and the responsibility for the safety of its passengers.

I got down on my knees and peered under the engine. The bottom of the sump was black and shiny with leaked oil.

When I arose the chauffeur was looking blank and it dawned on me that even if he could drive competently, given a sound vehicle, he knew little of cars as machinery.

"We'll be lucky if the engine isn't seized," I said.

I explained the position to my guides, who looked baffled and alarmed. Then the chauffeur explained things to them in Indonesian. They looked more baffled and alarmed.

"Do you know if there is some place near here where we could get a couple of gallons of engine oil?" I asked.

After a muffled discussion, which included the chauffeur, it was decided that they did know. I advised a telephone call, which was made from a nearby roadside store. About half an hour later, while we all stood, bored, by the car, a ramshackle truck arrived and its driver produced a can of oil from which he filled the engine. He spoke to the chauffeur and confirmed that the sump had, indeed, been absolutely empty.

"Well," I said, to cheer up my now disconsolate guides, "it's taken a few hours for all the oil to leak out. And perhaps it wasn't full when we started out. If the engine hasn't seized, it will probably start again and we can carry the rest of the oil with us to replenish the supply if we need to."

They, and the chauffeur, appeared reassured and looked at me with trusting expressions – which embarrassed me.

The chauffeur started the engine, adjusted his peaked cap, and we went on our merry way.

Chapter 38

Finally, after what had seemed a day and a half of essentially purposeless activity we had, like a randomly circling osprey, at last zoned in on our target. This was a relief, but the tambaks, when we reached them, were eerily like what I had imagined. Even the collection of small huts that formed a little village on a place where a number of the tambaks' retaining walls – built of excavated, dried mud – came together, resembled precisely those in the photographs I had seen in the Schuster report. In the twenty years since he had written, the city of Djakarta had spread its edges out so far that the tambaks were now right at one of those edges. When you raised your eyes, instead of palms and more tambaks, there was a formidable backdrop of city buildings – almost at touching distance, it seemed. So there was a jarring note, but that was not Schuster's fault.

There was a second jarring note: I quickly sensed my guides were not content. Indeed, they now looked and acted tense, depressed. We stood by the tambaks, saying little. I tried to get them talking. This, after all, was supposed to be their field of operations. Here were the tambaks, famous among aquaculturists the world over. It was my guides' moment to expound, mine to shut up and listen . . .

But they didn't want to talk about tambaks and their silence compelled me to ask questions. All I got as replies were grunts, or muttered, evasive stuff . . . or just more silence.

Schuster had stated that the socio-economics, even the culture, of the pond operators and their families, was dependent on the village-like concentrations of dwellings among the tambaks, connected to each other by a maze-like intricacy of pathways along the embankments between them. I asked if this was still true. But the replies I got were inconclusive.

I asked if it was still true that, as well as counting on at least one cash crop – and sometimes two cash crops – of fish a year, the tambak families could meet their own dietary needs for animal protein from milkfish production. And I asked whether it was still those people's custom to consume shrimps produced in the ponds, eat the eggs of birds that nested in the mangrove trees that edged the tambaks, and burn mangrove wood as a cooking fuel.

To none of my questions could I get a satisfactory answer.

Then it dawned. I knew as much, or more, about the tambaks than my hosts. I felt extremely embarrassed, at a loss for what to say next.

*　*　*

My task, as defined in the documents IEC had accepted from FAO, was supposed to be an evaluation of the efficiency and effectiveness of the tambaks as producers of food, and suggestions on how this could be improved. But I could now see how much more difficult this might be than FAO and IEC were anticipating. Any measure aimed at altering the day-to-day functioning of the tambak system, with the intention of increasing its fish production, would be ridiculous, if viewed in isolation from its human consequences. A significant change in the value of fish as a cash crop would not necessarily benefit the tambak operators. Any real and lasting benefits would need to include genuine improvements in conditions for the labour force, and in the network of social interactions and village life.

In short, my task was daunting to me. I lacked the self-confidence that could have quelled the shrinking inadequacy I felt.

*　*　*

Soon, two more young biologists from Djakarta, who had been invited to meet me by my guides, joined us. They were just as pleasant as the Toorang pair, but had nothing to add and, after half an hour of routine chat, the four of them stood talking quietly together in Indonesian by the side of a tambak while I took some photographs. The depression I was feeling was deepened as I finally dragged it out of my four companions, that the clearly neglected tambaks we were seeing were typical of the remains of the quarter million hectares of tambaks that had once been the pride and joy of fish culture in Java. No one was doing research on them. Indonesia, that Indonesia of the past, when a rural people had brilliantly developed a tambak industry through the undoubted peasant ingenuity Dr Natsir had referred to and that was entirely consistent, and in harmony, with their way of life, was gone. And I wondered why – and whether – tambak fish culture was now a lost art.

* * *

Back in the car half an hour later, headed back to Toorang, my two guides became polite but persistent questioners. I realised that this was the mission with which they had been charged – to pick the foreign expert's brains. This was unfortunate for them, because there was so little there for them to pick.

"Dr Logan," Mr Saleh began, "could you . . . ah . . . perhaps suggest ways by which the tambaks could be made . . . ah. . . more productive?"

Well, I thought, here it is at last! The *only* question, after all. But so meaningless, so simple. So huge. What did they really think I was? And surely they knew no "answer" I could toss off could really be worth anything. Didn't they know that? I thought they probably felt as miserable and embarrassed as I did.

I considered. I was a scientist of the West, and in those days still found it hard to oppose most forms of technological "progress". But now I was pleased that I could think of no prospect of a sudden, wrenching change of practice that might be applied to the operation of tambaks. If I could have, I might have been strongly tempted to view its consequences from a neo-Luddite perspective.

I don't think my reveries lasted more than twenty seconds of real time, though my guides may have been puzzled by a pause of even that length. But I was trying to imagine what I could do if I were, like either of them, an Indonesian fishery biologist without money or resources for major research.

I fixed on the tambak food supply. "Perhaps," I said, "there may be some way you could boost the production of algae for the fish. I mean, if you could discover the optimum concentrations of phosphorus and nitrogen for algal production, it might be feasible to increase it at low cost and small effort."

My guides looked perplexed, baffled.

"Even a twenty percent increase in fish production over the year could be quite valuable. You could do simple, inexpensive experiments on this. You could – "

I felt the uselessness of it. I was a scientist of the West. What they wanted from me was a wave of the technology wand. They, like Dr Natsir and Dr Hakim, had both studied in North America. They had seen how "we" did things. They wanted our conjuring tricks. But the stage on which the conjuring tricks were done was made of money. They had no money, and I was not the person to tell them how to get it in the quantities they would need.

I made a weak try to sound helpful. I said, "If you could get the milkfish to produce eggs in captivity, then you could eliminate that messy business of catching the young in the sea near beaches. That would save something, though I don't know whether . . . "

This produced hesitant smiles. "Maybe you could get the females and males into breeding condition whenever you want, by hormone manipulation," I said.

Now Mr Hamzah and Mr Saleh were nodding. All smiles.

"But," I warned, "then you'd have other problems to consider.

"First, the breeding stock would inevitably represent a very small gene pool compared to that in the sea population from which you collect eggs at present. Who's to know that the health of the fish in the tambaks isn't dependent on that large gene pool? I mean the genetic variability it represents. If you lost that you might be in lots of trouble."

The looks I was getting now weren't encouraging. They didn't want to know this.

"Then, you need to think hard about the social and eco-nomic life of the tambak network. Remember that Schuster – "

Mr Hamzah and Mr Saleh didn't want to remember Schuster. He was yesterday's man. The West would not think Schuster's way. Mr Hamzah and Mr Saleh wanted to think like people from the West. I was not talking like someone from the West was supposed to talk – all urging and enthusiasm and optimism. They wanted dreams, and the power to make them real, not a cold splash of reason.

Soon I fell silent, listening to the struggling engine of the BMW, absorbing the blows to my coccyx as we pounded our way over the potholed roads back to Toorang.

* * *

I left Indonesia a couple of days later, but during those days I spoke to several more biologists. What I got from them – often by inference rather than from direct statements – confirmed my own guesswork about what had happened to the tambaks. First had come the Dutch, who had exploited the country for its resources and its pool of cheap labour. The war and occupancy by the Japanese had followed. After the war had come national liberation battles ending, after a bloody struggle, with the removal of the Dutch, and after that more struggles among the Indonesians themselves as they quarrelled over different visions of self-rule and the power of government. All of it a generation-long series of deadly and draining upheavals inevitably inimical to any settled cultural, economic and scientific development of Indonesian society.

There was no doubt of the innate abilities of the people. Their strong, indigenous cultural past, the very fact that they had invented the tambak system and the country's agriculture, proved that. And I kept on meeting biologists, in addition to Dr Natsir, who had obtained overseas degrees, including PhDs, from major American and European universities. But the models of progress they had adopted were, I thought, wrong ones for Indonesia, models that led to stultification because they were being applied in a country that lacked the sort of intellectual history, the technical traditions and the essential wealth required to employ them. That was why the progress in tambak design and operation was – if not dead – suspended.

* * *

My report to IEC, which Rudy Hauser would send on to FAO, was a low point. Rudy, for the first time since I had known him, was visibly disappointed with what I had to say.

"Surely there must have been some way you could have suggested things, or helped them set up experiments so that – "

"Look, Rudy," I said, "with respect, you weren't there. Down the road from Toorang, American interests are setting up a huge soft drink factory. There are active plans for more high-rise hotels in Djakarta. And other foreign operators are financing – for God's sake – a new race course. Everything's Western stuff. And unlike us – I mean those of us who think we've got the scales off our eyes – it looks like too many Indonesians aren't interested in the sorts of modest solutions to their problems that would be consistent with the true state of their present economic and technical capabilities. I mean solutions that might lead to solid, permanent growth over the long haul. What they wanted from me – as innumerable other foreigners have persuaded them they have a right to expect – were suggestions for a technological quick fix."

After a while Rudy nodded. "Okay, Bill. You may be right. But FAO won't like it."

"Which is their problem. From what I can tell FAO have lots of people in the field in Third World places, some of whom are indulging in damagingly Western thinking, or seem incapable of encouraging the locals to make practical and valuable adjustments to their existing production systems. In many instances, those production systems have functioned damned well, having been worked out over the ages to fit their particular environmental conditions and the needs of the population. My report says all this in relation to tambaks. Maybe it'll even do a bit of good if it can inject a tiny dose of reality into what can easily become a very flatulent discussion."

"In short, you want them to develop – if at all – without foreign aid."

"Given that foreign aid always seems to come with unholy price tags in the long run, as far as the ultimate effects on financially squeezed societies are concerned, and is usually related to the maintenance of Western hegemonies in trade and distribution and profits, of course I do."

Rudy looked at me steadily, if a bit uncertainly.

"You really aren't very American in your thinking, are you, Bill?" he said, at last, with an ironic chuckle.

I shrugged.

"I just try to call things as I see them," I said.

Chapter 39

Betty had expected me to be away for less than the time she would be in Bolivia. As it was, I was back two weeks after I had departed from Australia.

A few days later I received a letter from Andrew Peacock. The part that mattered said, "From a lot of things you left unsaid, or only partly said, it has slowly dawned on me that I may have frightened you off the turtles because you thought I had prior claim. If so, I feel badly. IEC sent you here at the request of the Australian government and FAO to study and report on turtles. There was no way you could have known I had started work on them a year and a half ago, because hardly anyone here was much aware of it. I was in the field a lot of the time and hadn't published anything. Anyway, Bill, to cut it short, having met you and your more than charming, and beautiful and . . . well, rapturously lovely film star wife, I want to ask you if you can possibly come back here to work as my colleague – equal shares – for a year or two. I know next to nothing about how IEC manages its business, but I feel you and I could get along. I've looked up some of your papers and I'm particularly taken with the coyote stuff. I'm positive we could do some good work here together. And let's face it, I need a colleague. Too much here for one person. But I only want someone I feel would do a good job. It seems clear that you would, and I would be more than delighted if you could swing it – I mean come – if it would be to your liking. "I feel an utter idiot for not saying these things when you were here. Perhaps I was so bedazzled by Betty that . . . Just joking. Anyway, I eagerly await your response."

I showed the letter to Rudy, who said, "Yes, it certainly is a pity he couldn't have said this when you were there. Would have saved us quite a lot in air fares. Of course, you'd still have had to go to Indonesia, but then you could have returned to Australia from there."

"True," I said, "but perhaps I was much too hasty in assuming he wouldn't want a co-worker."

Rudy nodded – a bit absently.

"The thing is," I said, "while knowing how Americans tend to think about these matters, it seems I still retain a measure of Anglo-Australian attitude about how one views one's own, and other people's, scientific turf."

"Meaning?"

"Meaning that I tend to say to myself 'hands off!' if someone else has staked a claim to a research problem before I get there."

Rudy laughed. "You mean, if you were a 'bloody Yank' you'd just slot in wherever you could and try to outcompete the other guy."

"Crisply put, Rudy. Very crisply put." We both laughed.

"Look, Bill," he said, suddenly collecting himself, "go for a year, see how it pans out. Maybe go for two years. I think we can fix that. That is if you feel as trusting towards Peacock as he seems to be towards you. Do you?"

"Sure," I said. "Certainly. I think."

"So go. This turtle project is of great biological and conservationist interest. We ought to have a share in it as we originally intended. So let's do this, if you think Peacock won't end up resenting you. Cable him your acceptance. No, wait. All very well for me to say. But what will Betty think? We're talking about a long time away."

"I'll phone her in La Paz. I think they're shooting near there. She won't be back in the States for at least three months."

*   *   *

"How's it going, love?" I asked.

Her voice came back over a good deal of crackle on the line. "Oh, shooting problems, Bill. The Bolivian crew are not the easiest to deal with because their contract has them working for a quarter of the dollars Americans get for the same job, and they're understandably bitching about that." This view, I thought, indicated a certain liberalization in Betty's political thinking since the shooting of *Tono-Bungay.*

"Oh," she went on, "and the producer's having tremendous trouble getting enough uniforms for the battle scenes."

"Is it fun?"

"Fun? Not the right word, really. Oh, it's a mighty challenge, and eventually it might be a mighty film. But no. Not a lot of fun at the moment. How is it with you, darling?"

I explained what had turned up; she was silent.

"For God's sake, Betty," I said, "don't go quiet on me. Shout abuse if you like. Tell me you hate me . . . or anyway that you hate this idea."

She spoke very quietly, though. I could hardly hear it above the crackle of the line. "The trouble is, dear, I can't think of what to say. I feel I need you now. But you have a job, too, and can't come. I know. And when I do get back it'll be for a lot more shooting in the studio. And if we get together it'll only be for a few days now and then. Bill, can we . . . can we pull this thing off?"

"Let's say so, love," I said. "Let's not speak of alternatives."

There was a silence that lingered.

"Are you there?" I said.

"Yes. Just."

After another pause she said, "Look, Bill, of course you'll have to go. Andrew Peacock seemed like fun. And the turtles were marvellous. It would be great for you. I just wish I could be there, too."

"But . . . you still want to do films, don't you?"

"Oh, I suppose I do, but . . . "

It wound down after a few minutes more. We discussed how she might – again – join us for a while after the Bolivar film was in the can. If she had the energy left by then to come to Australia.

*   *   *

That night I wrote a letter. The last part was the most important.

Darling, for the first time I've sensed that you truly might not want your film work to last for too many years. I've been careful to keep off this subject, because you've seemed to want to do it, regardless of the undoubted stresses, and because you're tremendously good at it. That I believe you're tremendously good at it might, I suppose, be not completely unexpected. But others in the film business have told me that you're widely viewed as the best actress in movies under the age of forty, or something like that. I don't believe these people were joking. And reviews I keep seeing appear to confirm their judgement. In the face of such opinions how could I – a humble biologist – prevail on you in any degree to change your plans or the trajectory of your career? Unless I sensed in you some sort of sign that you just might not want to pursue that career – however glamorous and rewarding – forever.

For me too, just now, my life is concentrated on my work. And I'm under no stress in believing I want to stay with it. The only stress for me is our separation. Try and think very hard, Betty, what you want. The prestige and money in movies are very nice. But peace of mind and fulfilment should be the first considerations. If either of us lacks them, no other thing is worth the price, in my view.

Chapter 40

Rudy and I finally arranged for me to be at Sydney University as a base. One reason for this decision was because I already knew Larry Marlowe from the Zoology Department, who had accompanied me on my initial visit to the Barrier Reef, and we were familiar with each other's work and got along well. Another reason was that, although Sydney was a long way from the Reef, Rudy wanted me – as part of an IEC global evaluation process – to view and compile reports on the work of a number of wildlife and fishery biologists and conservationists in Australia. For that, Sydney was the best placed of Australian universities. For this evaluation work I would need to get to know a number of people and see their work in laboratory and field wherever I was welcome. I would then invite them to contribute chapters in a book Rudy and I would eventually edit. We planned to write a long "wrap up" chapter ourselves, in which we would discuss the personal viewpoints and future workplans of the authors. Rudy's idea was that IEC should put out volumes like this for Australia, New Zealand, and for some parts of the Far East, Africa and South America. "All the places," he said, "where you have vast spaces still not too completely modified by human hands, plus tremendously interesting animals and plants with not much known about them yet. A set of volumes of this kind would let us see

where the approaches differ regionally and where they converge –
and why the differences and similarities exist. And also where the
international funding of science – when we can get any – could be
best put, so far as the future of wildlife and fisheries and ecosystem
research are concerned."

I said the only thing wrong with this idea was that it would
take ten times longer than Rudy was suggesting . . . or ten times the
number of IEC personnel. Rudy grinned. "Okay, maybe you can't
do it. Not if you're going to spend a lot of time with Peacock. But
don't forget that you can't do much on the turtles for more than half
of each twelve months, because four months are all the time they'll
be nesting. You can at least make a start on creating a liaison with
the people working on kangaroos and various bird species – I hear
there's some great work being done on magpies – and the fishery
workers. You actually know a lot of the fishery people, don't you,
from your days with FROA?"

He smiled at me then. "I believe your old friend Percy
Swanson is now Head of the Zoology Department at Sydney."

I knew that. Swanson had been overseas when Larry
Marlowe had accompanied me to the Barrier Reef.

"Swanson's getting to be a big name, by the way," said Rudy,
still with a little grin.

"I'm aware of that, but he's a very different kind of ecologist
from you and me."

"Well, that's okay," said Rudy. "Good thing in fact. Everyone
needs a dose of the other guy's medicine. Otherwise we just become
polemicists for our brand of scientism. Running into major intellectual
opponents saves us from too much navel-gazing and self-satisfaction."

I agreed with him, while thinking that this attitude was not
too common among eminent American scientists that I had met.

"But," I said, "don't you remember what Swanson can be
like?"

" I haven't seen him for quite a few years."

"Well, he's a tough polemicist. Whatever you say to him
you're apt to get a strong argument."

"Sounds like fun."

"Fun be buggered," I said.

Three days later I was in Sydney.

* * *

I ran into Percy Swanson in the main hallway of the Sydney Zoology Department almost as soon as I got though the front door. He was probably on his way to lunch in the University Union, but he stopped to talk. He seemed friendly enough.

Anyway, he had read my research reports on the coyotes and the thermal spring. "Ah, yes," he said, quite amiably, "they were interesting work, those things. Particularly the thermal spring work. What was the name of it? Something about –"

"Hell," I said. He opened his mouth and frowned. "Hellhole Springs," I said.

"Of course. Hellhole Springs. Very colourful name."

I didn't know whether he was the same Percy Swanson of years before. He had expressed neither dismay nor delight at seeing me. Impossible to divine whether he loved or hated the idea of one of his former students – one with rather upstart attitudes – returning from years overseas, not exactly covered with glory but with some sort of modest achievement. But it seemed hardly the time to worry about that; after all, he had been responsible for my getting to work with Rudy in the first place.

I soon found out how little he had changed. Ten minutes into our meeting and he had still not gone through the door heading for lunch but instead was saying that "Hauser has a considerable reputation in America, I know, but to be honest I'm not entirely sure I understand what it's based on. I can't believe that really great advances in ecology are going to come from his way of approaching the truly fundamental questions." From the tone of his voice and the hint of battle in his eyes, the clear implication was that his – Swanson's – was the right approach.

But I didn't feel much like a polemical fireworks display so soon after arriving. Instead, I thought of an artful dodge that happened to be based on fact.

"Rudy Hauser is actually an admirer of your work, Professor Swanson."

But if I thought that might be a sop to his ego I had underestimated him. "Is that so?" he said dryly. "Well, then, I have to wonder why he doesn't modify his own approach, because I don't see how he can truly 'admire' my work – which undercuts his own in concept and method – unless he is prepared to change his own veiws accordingly."

I knew I would soon get into a very messy argument with this man who was a born controversialist, so somehow – I forget the details – I found some excuse to wriggle out of further talk.

The light of conflict died in Swanson's eyes; he would have liked a real verbal brawl, but I wasn't going to give him a fight just at that point. If it had to happen, I wanted time and leisure to choose the place and the occasion.

A complication was that I was not altogether opposed to Swanson's arguments. I had adopted his approach, in part, in the Hellhole Springs study and in the Dixon Creek work. Both had called for lots of field work, but also experiments to interpret field phenomena. I was a scientific pragmatist, so far as getting a problem analysed; always prepared to adopt a blend of approaches – the reductionist experiment and the holistic ecosystem concept equally convenient to my mind – their use defined by the conditions imposed by the problem.

Of course, it was easy to like Hauser as a person and admire him as a scientist, and very difficult to do more than respect Swanson. But I had not forgotten the sharpness of Swanson's mind. "If you get into an intellectual fight with him," I told myself, "be prepared to be hit hard, and to hit hard in return. This guy's no intellectual lightweight. He'll hit you with everything, and from every angle."

Chapter 41

Andrew Peacock, as a hard-bitten Yorkshireman, usually spoke his mind . . . unreservedly. I was more accustomed to conceal my feelings, but in his presence found myself responding in kind. We managed well, though, day to day, splitting work and responsibilities right down the middle. We were both young and very strong and could work tirelessly over long hours.

As evening approached we would begin each night's turtle watch, roaming the coral sand beaches of the tiny elliptical cay where much of Andrew's research was being done. For hours the turtles would come ashore. They were not hard to see, even if there was no moon; the soft tropical starlight was enough. But they were readily turned back by sight or sound of us. We had to be careful, as they crawled in, to include in our nightly counts of turtles only those we were sure were going to continue up the beach until they reached a suitable site for egg-laying.

For the work, we had to get established on the beach every evening before the arrival of the first of the night's nesting females. A major aim was to obtain a reasonable estimate of the total egg production of all the turtles that laid eggs on the cay in a season – an endeavour complicated by the fact that a turtle could lay several clutches of eggs in a breeding season. We were already familiar with the sight of turtles digging egg chambers by means of their hind flippers. Our task was to reach down under each turtle's short tail into the egg chamber and collect every egg deposited, counting them into a plastic bag. As soon as the last egg was laid we would rapidly replace them in the chamber, which every turtle would always cover with sand in a normal manner. In a few instances we retained the eggs for weighing and lab experiments. Even then, the female would pull sand into the empty egg chamber and fill in its excavated body pit as usual, and we would shake our heads, pat the turtle affectionately on its shell and say something like, "Stupid bloody animal."

While egg-laying was being completed and before the oviposition site was restored, we used flashlights to make a number of measurements on each turtle, including body length.

After collecting sets of body measurements and corresponding egg counts on more than a hundred turtles, it became obvious that egg number was a function of body size. Turtles deposited an average of about a hundred and fifteen eggs; the smaller turtles laid about eighty-five eggs and the bigger ones about a hundred and forty. This was the broad trend; a few turtles laid only fifty eggs, one of the largest laid two hundred.

After each turtle laid its eggs, before it could move off to return to the sea, we would attach a metal tag to one of its flippers so that returners could be identified, and by successive egg counts the total production of each turtle, and eventually the total for all the turtles visiting the cay, could be determined.

Green turtles were known to possess the potential for long life, though the great majority of hatchlings perished rapidly in the sea. For those very few that survived to grow large enough to lay eggs – estimated at about eight years and eighty-five centimetres in length – it was believed that life could continue until the turtles were more than twenty years and measured a hundred and twenty centimetres. Andrew believed it would be a quarter of a century before he had reliable quantitative information on turtle longevity – partly obtained by direct recovery, or return by others, of the metal tags. He also insisted that if age, infirmity, or "something else" prevented him from tracking these tagged turtles over extended time, then "some other bugger will do it and be thanking me for having tagged them now."

He also hoped that the tagged animals would reveal where and how far the turtles migrated away from the Reef after completion of their seasonal egg laying, and how frequently they returned to lay eggs in later years. Early reports of the turtles he had already tagged indicated movements exceeding three thousand kilometres in the directions of New Caledonia and south-eastern Papua.

Soon after I joined him we embarked on two problems: determination of the conditions and properties of beach sand that turtles selected for their nest construction, and how females located islands on dark, moonless, overcast nights.

We noticed several successful nests being made where the beach slope suddenly steepened, so we proposed a tentative hypothesis that Andrew explained, in his usual trenchant fashion, to a group of visiting senior students from the University of Melbourne who were doing a course in tropical marine ecology: "It seems to us that the poor bloody turtles would be practically clapped out after crawling on their bellies over the beach rock and sand, and would be absolutely pissed off at the prospect of dragging their miserable arses up a steeper slope than they had to."

The students were grinning a bit stiffly. They were unused to such speeches from lecturers. Andrew just grinned back at them. "If you're wondering why, if they're so bloody tired, they don't stop almost as soon as they get out of the sea – I mean, why they come up the beach as far as they do – just remember the eggs are porous to air. They get oxygen, as they develop, from the air that reaches them through the interstices of the beach sand. But if the eggs aren't deposited safely above the highest tide level, they'll get inundated by sea water and die from oxygen lack."

A bit later, when we noticed some nests constructed just at the point where the cay's vegetation of pandanus palms, casuarina and pisonia trees began, we thought of a second hypothesis – that perhaps the onset of vegetation was also a target for turtles to aim at and a trigger, when they reached it, to release digging activity.

Peacock had brought two students with him from the biology department of the University of North Queensland, and we invited the Melbourne students to join us. For a few nights we engaged in some wild fooling around with the sand slope and vegetation hypotheses. It all seems pretty ridiculous to look back on, though I never had much more fun in doing field work. To test the sand hypothesis we handed out shovels to four students while others were set as spies around the cay's perimeter to report when turtles were well up the beach and advancing purposefully. Then we would hurry to a position in front of the approaching turtle and there have the "shovel detail" construct an instant slope of sand. We thought, if the beach slope idea was worth a damn, that the turtles should show at least some slight interest in trying to dig where our artificial slopes began. However, these instant slopes never gave any sign of being successful. The turtles either adopted a new line of approach up the beach to go right around them, or else they were scared off and went back to the water.

To test the second hypothesis we spent hours building a wooden frame to which we nailed branches of pandanus and casuarina. Then, at night, with a squad of half a dozen of the sturdiest male students to lug this heavy, awkward, four- metre- long screen around, we all kept watch for beaching turtles. Again, the students who were not screen movers were deployed to report on beaching turtles. Once a turtle was moving firmly up a stretch of beach our screen movers would trundle their screen along the beach and attempt to place the frame at right angles to the turtle's projected line of progress. If our hypothesis held, the turtle was supposed to come up to the screen, then stop and dig. But many of the turtles deviated from their original course up the beach so that the students, like rapid scenery changers in modern open stage theatre, would shuffle rapidly with the screen, to place it across the new line of approach. Such moves spooked many turtles sufficiently to turn them back to the water. Sometimes a turtle would start digging well before reaching the screen. Sometimes a turtle would pass very close to the screen, even brushing it, but never stopping by it as in that one earlier experience on my very first visit to the cay; instead always moving on further up the beach to dig. Well . . . we had a lot of fun for several nights, though I don't believe the students thought we or they were behaving with quite the gravity appropriate for real scientists.

A week or two after these comico-farcical tests of our hypotheses, they were thoroughly destroyed when we observed a number of turtles that climbed the steepest sand slopes on the cay, and then passed under and beyond the palms and casuarinas to make their nests. So we never did solve the question of what released digging activity.

One thing we were able to nail down with a degree of certainty was the role of sand qualities in the building of successful nests.

Andrew had already noted that nests made in dry sand were

apt to cave in while under construction, leading to their abandonment. In sand that was too moist, the difficulties of excavation also resulted in abandonment. And, where nests were dug too close to the bases of trees, an abundance of fine rootlets interfered with the digging.

We studied success in digging, scoring it in relation to the sand factors that we named "dry", "normal", or "moist", and "no", "few" or "many" rootlets. We found the optimum conditions for making good nests occurred in sand of "normal" moisture and "few" rootlets. Such sand was easy for the turtles to excavate yet held its shape in construction of both body pit and egg chamber. In dry, rootless sand, even when nests were successfully constructed, the eggs tended to become desiccated, whereas moist sand, if it could be excavated at all, increased the chances of the eggs being smothered because of air being unable to reach them.

In a further study, a temperature sensor was located in the middle of an egg clutch, another at the edge, a third two metres further away in the sand at the same depth, all three connected to a continuous recorder. Just before hatching, following a development period of about two months, the temperature at the centre of the egg mass reached a value six degrees Celsius higher than that in the sand two metres away. This rise was due to the energy expenditure by the developing young, as they consumed yolk during their incubation.

We found, in a lab experiment with turtle eggs we stole from the mother just after they were deposited in the egg chamber, that those held at thirty-three degrees Celsius hatched in forty days, whereas those held at twenty-seven degrees required twice as long. The appearance of the hatchlings at the lower temperature was subtly different from those at the higher: they were larger and looked somehow less juvenile – more differentiated or "mature" – in their body form. This study had been inspired by the knowledge that our cay was at the southern limit of the Barrier Reef turtles – the coolest environment in which they deposited eggs. It was known that eggs held at twenty-four degrees failed to develop. It was also clear that eggs laid further north would develop faster, and the higher temperature we used was similar to the sand temperatures found there. We were not sure how to use the results of this experiment at the time, but believed it would help eventually in understanding turtle ecology over the entire southern hemisphere range of the species. It could, for instance, be ecologically important that in the warmer tropics the developing eggs would be in the sand much less time.

That way, the eggs would be less exposed to predators such as rodents, which dig them up for food.

I wondered about the functional significance of the soft, flexible shell on the pingpong ball-like eggs. Its porous nature clearly enabled the developing turtle to respire. But since it was porous to air, could it function also in water conservation, supposing that eggs were exposed to desiccation in a particularly dry area of sand? I performed a simple experiment to test this.

Starting with whole eggs that were still covered with mucus from the mother's birth canal as controls, I carefully removed the mucus from a second group of eggs and, with a third group, I dissected off the entire top third of the egg shell. I then weighed all the eggs, following which I exposed them to identical conditions of temperature and low humidity and weighed them at fixed intervals.

All the eggs lost weight, due to evaporation, at essentially the same rate. So eggs with an intact shell, whether mucus-coated or not, were apparently no more effective in preventing water loss than eggs with a third of the shell removed. The shell, it seemed, was no more than a porous container for the egg. It was the egg chamber environment that determined whether or not eggs would survive their development period. You could see, with added emphasis, how important it was for the turtles to select sand of a satisfactory moisture level if eggs were not to desiccate rapidly.

*   *   *

We never found definitive explanations of how turtles located the cay at night. But we did wade out far over the reef one very dark night, and then extinguished our lights. While feeling this was not a good time to be on the reef, up to our chests in water, exposed to whatever might be swimming near, we were surprised to discover that we could make out the beach as a sort of luminous line, and the mass of the cay's vegetation as a band – darker than the moonless night sky – above it. Years later someone reported formal experiments that seemed to confirm our simple observations: that turtles, like humans, possessed the ability to distinguish the form of an island even on dark nights.

Explaining the incomparably more impressive feat of navigation by which turtles could find their way back to a minute island after a year or more's absence – which was what many of them could do – was far beyond the scope of our observations.

*   *   *

We did these things in the period from the end of November, when the mature females first hauled out of the sea to lay eggs, to the end of February, when egg-laying ceased for the season. Our records of nightly beachings showed a distinct cycle of numbers that peaked five times at intervals of about two and a half weeks, and coincided with tides.

When the season ended I returned to Sydney for a brief vacation, and to catch up on correspondence from the States. Andrew went to Brisbane. We were to meet in three weeks to analyse our results and decide on next steps. We were quite pleased with the amount of useful work we had been able to complete.

Chapter 42

My overseas mail came each week by launch from the Queensland mainland where it had been forwarded from Sydney. I had received no letters from Betty for several weeks, and was wondering . . .

I had written asking the secretary of the Sydney zoology department to hold all my mail during the last three weeks before I was scheduled to return. A small pile of letters awaited me in Sydney. Still nothing from Betty, but one letter was from Los Angeles. It was addressed in a florid, heroic hand. The writing was not familiar to me, but I somehow guessed the author's identity.

"Dear Bill," it said. "Greetings to you Down Under. I have not visited your homeland, though I feel I must manage to do so before I am too old.

"I trust that the work you are embarked on is intellectually rewarding. I am conscious of the unlikelihood of its being financially rewarding. I continue, as always, to admire people such as yourself who are able to derive their satisfactions and fulfilments almost entirely from the nature and purpose of their work.

"Bill, I don't know quite how to broach with you what you will probably consider an intensely personal and sensitive matter. But I am going to do it all the same because, of all the young people

I have had business with in connection with films your wife Betty is – apart from being the finest screen actress of her generation – the loveliest and, quite simply, one of the nicest young women I have known in the movie industry. Apart from this, Bill, I have a high regard for you also, and have wished for you both that your marriage would be spared the problems that afflict so many of those connected with this business of film making.

"Okay, I know you, Bill. A blunt, direct Australian. I can hear you saying 'Come to the bloody point, mate.'

"Well, the point is this. Betty is back from Bolivia (perhaps you know this, perhaps not), and she is a troubled girl. All accounts I've had suggest she turned in a marvellous job of acting. She's brave. She's a trouper. People like her. As you would say, 'no worries' there. Or at least no worries she's let defeat her. Yet . . .

"But Bill, I've just seen her, and she looks drained, as if she's been under immense pressure. She's nervous and anxious. But all that aside, she's talked to me, and what is bugging her is that you two are a world apart. I have suggested, urged, entreated her to go see you in Australia, immediately. She may do that, but if she does I think you need to be prepared, Bill. For it seems to me, here and now, she will not be able to endure much more.

"Yet it seems as if she is not ready to quit the film business.

"And, most serious of all, she may be about to find someone who can relieve her discomfort. Bill, I can't put it more frankly than this.

"I do not mean she believes this person – a young Italian assistant director of the Bolivian film crew that she met during the Bolivar shooting – is a substitute for you, Bill. She has not, I feel sure, been unfaithful to you. But this man is now a close friend, rapidly becoming a very close friend indeed. She literally doesn't want an affair with him; she has told me that. But she does like him a lot (he is in fact a pleasant-seeming and presentable young man – not at all a blunt fellow like you!). And I think if things went the wrong way he is the sort of person who, because of the nature of his devotion to her, would be at her side. Yes, I know how you love her, but with you it's different.

"Well, Bill, I cannot draw you a much more detailed picture. I think too much of both of you to do anything but be concerned at this turn of events. I could say 'get your ass over here on the next plane if you want to save your marriage!' But I respect you as I respect her, and I know you are no more likely to drop your work than she is hers.

"An unredeemable mess, then? Jesus, Bill, I hope not. I'd like to say to you both: For Chrissake, kids, work something out. You had it so good. All of us could see that. Hang onto the ball, Mate! But bloody do something! Oz enough for you?

"Fond regards,
"Irve Robichaud"

*   *   *

Hard on the heels of Irve Robichaud's letter came a telegram from Betty herself:

Bill. Need to see you. Hope you are in Sydney. Otherwise this will be forwarded. Arriving Sydney 28 March American Airlines Flight 607. Love. Betty.

Chapter 43

She emerged from the passenger exit looking, as Irve Robichaud had written, drained. I knew from my bathroom mirror that I looked pretty used up myself. Nine nights of little sleep can undermine you quite effectively. But Betty's appearance hit me like a clarion call of woe. There, in the face of a young and beautiful woman, I could suddenly sense how she might appear in age and in the grip of a debilitating illness.

I felt an instant of pity and foreboding before it came to me how young she still was, and that what I was seeing could probably be banished by the reassurance that what she had been going through was not yet the end of the line. I felt I had only to stretch out my hand and take her hand in it and all would be well again and . . . I did stretch out my hand, and she did take it. And we hugged and kissed as we had always done in greeting.

And then it came smashing at me that this part was easy, but that it would assuredly not be all it would take to resolve matters.

And as we talked in low tones and searched each other's eyes, I knew, for the first time since we had met, the imminence of a doom that I had felt could never touch us. I mean that, with neither war, nor famine, illness, death, or allegiance to conflicting ideals to come between us, our oneness when together was disintegrating. If, in separation, our longing for each other had been strong and certain, it had always – in both our minds – been instantly assuaged when we came together. Now, even as we faced each other at no more than the length of an arm, only strands of the full dazzle of our attraction for each other were filtering through. I wanted her, and she wanted me, but the mutual taking – that had always depended on instant mutual acceptance – had something in its way.

Oh, our embrace was tender, but it seemed somehow like the hug of sympathy one gives a grieving friend rather than the hearty desperation of lovers.

Again came the feeling of age. I was only thirty-one, she twenty-eight, but I could suddenly taste the cold iron of eternity.

She didn't beat about the bush, though. Not in her. Not the style of either of us.

"I can hardly tell you this, Bill . . . but maybe you can even guess what I want – "

"Yes, dear. I think I can." Well, probably I would have been able to guess even if I hadn't heard from Irve Robichaud.

"It's not you," she said. "And . . . and it's not us. And really . . . it's not me."

I waited.

"And the devil of it is, I don't see how we can do a lot about it. I mean . . ."

"You mean," I said, "that you know I love you and you love me, yet it's not as if we can fix things by either of us just dumbly swearing to follow the other to the ends of the earth. Or that both of us will quit our jobs and retire to an island and run a general store."

Her eyes clouded and she shook with sudden sobbing.

I held her tight and fought my own tears.

"Damn you, Bill," she gasped, "for being there that night at the theatre for me to meet. Everything would have sailed along okay and I'd have been . . . well . . . okay. Not great or marvellous, but okay enough for me to have done my acting and – "

"I could pretty well say the same thing," I said. "But the real trouble is that neither of us is ever going to think the world was well lost for the other."

"What d'you mean?"

"I mean," I said, trying to keep my tone steady, even stern, even didactic, "that though it's a tough message to digest, we're saying that love isn't enough. We love each other, and until now I'd have mouthed the cliche: 'more than anything else'. But that's false, because neither of us is going to denounce work. We need it if the love is going to hold us."

She gave a slight half-nod, but said, "Yes, but work. Isn't letting it be so important very selfish?"

"I don't think so. Look, a generation ago I'd have worked at my profession and we'd have married and you'd have kept house for us and our three kids. That's not there for us. You and the world – and I – believe you have a right to work and that it's your destiny. But no one's going to say to me that *I* mustn't work because you're a great actress. In my own way, I was destined to do the work I'm doing, long before we met, just like you."

We were holding each other very tightly again and, to break out of this situation which, to me, with a growing sense of its irony, seemed to demand something mundane to crack it up, I said, "Love, it's half past two on a fine afternoon. I've been waiting around this bloody airport for hours. I'm hungry. And I need a drink. Let's grab some lunch."

Betty pushed me away, to look at me with an appearance of something between anger and hurt. But then despite anger and hurt . . . and tears, she found she had to manage a smile. Then a larger smile. Finally she laughed and embraced me.

"Okay, you Oz ocker," she said, still laughing and crying. "Lunch!"

* * *

I felt it could be getting close to the end of things though.

Chapter 44

It took forty minutes to drive from the airport to a seafood restaurant at Watson's Bay, a place just inside the South Head of Sydney Harbour, a quiet place because too far out of town to have many tourists this long after the summer season.

Betty, after a few nervous remarks and some unaccustomedly anxious smiles, had showed her exhaustion by dropping off suddenly into a deep sleep, like a child, as we sped across Sydney. When we pulled into a parking space fifty yards from the restaurant, I woke her with a gentle kiss.

The great eyes opened – eyes now known to millions – and I took her hand. "Lunchtime, love. Care for a nice piece of John Dory?" She smiled. Her old beautiful smile. She was still drawn and pale but looking young again.

We walked along the stretch of pavement to the plain wooden building that housed the seafood restaurant. Its windows looked out at a strip of sand beach and the Harbour. Through the windows we could see the old wooden wharf where the ferries used to call when they ran regularly on this part of the Harbour, years ago, and where that hammerhead shark my uncle had caught had hung from a hook for public display.

The restaurant was as unpretentious inside as it was outside.

But the afternoon sun of early May bathed it in glowing warmth, and the fish we ordered was perfectly fresh, perfectly cooked in the plainest way, the servings large. We ate slowly in appreciation of the food, the white wine that accompanied it, and the Harbour view of blue water and passing yachts. Betty, still feeling almost shaky after waking, gradually relaxed, got dreamy looking, put her hand in mine and talked in a rambling but not incoherent way.

"The movie?" I enquired.

"Oh . . . you mean 'Bolivar's Bullshit'. Yes. Well, I think it's going to be what they call a great wide-screen epic about one of the heroes of South American history. It will probably be a box office smash, for whatever that signifies. The trouble is you can have truth as to story details, actors made up to look like those they represent historically, mouthing lines their characters are supposed to have spoken historically. And locale, scenery, wardrobe, and so on can have complete verisimilitude. And it's still sort of a farce."

"Because?"

"Because the producer's on contract to a big studio. Because some actors are ignorant slobs. Or drunken bums. Because no one has a real historical sense. Because they don't have truth in their souls."

"And the story, the screenplay?"

"Great story, good screenplay. A few decent actors. A director who used to be good, but who's tired and cynical. A Bolivian company that's young and quite inspired but who are put down by some Hollywood bullies who think they're God."

"But the result won't be a flop?"

"Let's say it'll be described as a competent, big budget, money-making pic, that serious critics will pan for its lack of an artistic soul."

"You're cynical, Betty."

"Sure. But my contract will be up before too long. Then I can start to look for stuff that's worth doing, and people worth doing stuff with."

"Would you go back to the stage, then?"

She hesitated, looking at me. "That'd be better for us, you think?"

"I don't know."

"Well, it might be. Better, I mean. Better for me . . . better for my soul. But I'd probably have to go to the east coast. And what would you do then, Bill? I mean, your own base is in San Francisco at the moment. And later, well, maybe you'd like to come back to Australia. What do I do if you decide to do that?"

We were still looking at years, it seemed, before anything was likely to shake down for either of us into something resembling a pattern.

"Do you *like* film acting?" I asked. "I mean you've said you'd probably always prefer the stage."

She looked out at the Harbour light. "Films have their fascination if they're done right. You get to leave something behind that people might still want to look at fifty years later."

"And that's important to you, I suppose."

"Well, yes. Actually it is."

We finished in silence and sat a few minutes over coffee.

"It's wonderful here," she said at last. "Real peace. Being in Oz is still like jumping off the edge of the world, almost like falling into a simpler, brighter, warmer universe. Will it remain like that?"

"Maybe in our lifetimes. Not in the long run. Everything changes. More's the pity."

"You feel that?"

"Well, I do and I don't. Nothing can be more smugly self-assured than some Australians. I don't mean they're arrogant as some nationalities can be – I won't say who I mean." I looked at Betty, grinning. "I know in my bones I'm a provincial. Wherever I live and whatever I do, I'll always be that. I can live in the big world but I have the provincial's mistrust of it and all its works. It's all Sodom and Gomorrah, Betty."

We laughed. "I wish to God we'd met here, Bill," she said, "while we were still kids, or maybe in San Diego, and settled down to some nice local jobs."

"You mean me as, say, a suburban dentist, and you as my nifty little devoted housewife?"

"Cut it," she said.

"Love," I said, "we're just little cogs in a great, gaudy historical machine. Tiny cogs without comprehension or cognition. Cogs without cognition! That has quite a ring to it. I could lend you a book that –"

"Cut it!" she said, as we both laughed in low voices and I paid the bill and we got up and walked out under the high blue sky in the glory of the Harbour sunshine.

* * *

She stayed for nearly three weeks. We drove a car out west to see the great grazing and wheat lands of New South Wales, out to Narrandera on the Murrumbidgee River. We saw sheep in millions, and red kangaroos, flocks of hundreds of white cockatoos and galahs. We looped north on our way back to Sydney, and passed through emu ranges. A few days later, after lounging around Sydney Harbour, we dipped down the south coast, through the tall posts of the spotted gum forests with their strange understory of primitive cycad palms, and through a narrow band of rain forest. In some of the giant trees there we saw domestic fowls that had gone wild and were capable of fairly strong flight. I wondered if I was like one of them. They sat up in the tree branches, sixty feet above the ground, leaner and stringier, and far more athletic than their barnyard counterparts, but as I said to Betty, "still unmistakably chooks." Yes, perhaps they were like a parable for me . . .

We came out on stony headlands and lonely beaches, with grey kangaroos hopping tamely out of the scrub every evening.

It was early May and the sea was getting cooler, or anyway, the winds that blew off it were getting cold. But on a few hot afternoons we plunged in and rode the surf for a couple of hours. Not every beach was safe and once we battled in an undertow for twenty minutes, in water not much over our heads and not more than twenty-five yards from the beach, before an exceptionally strong wave tumbled us unceremoniously, exhausted and cold, on the sand. We looked at each other with silly grins. It might have ended there, just out of reach of a long, deserted, gleaming beach, with the dark, unbroken scrub behind it, in the blue water rucked up by the wind all the way out to the horizon. Above us were huge, indifferent anvils of white cumulus.

* * *

In the end, nothing was solved. In her honest way that seems to have been so rare in women of that time and was what I prized her for as much as for her beauty and her spirit and brains, she spoke of the young Italo-Bolivian director and how he felt about her.

"He says he loves me, and I guess he means it."

"And you?"

"I like him a lot. I could never love him as I love you, Bill. He's a nice guy. Gentle, undemanding, and . . ."

"And?"

"And he's there."

"Well, but won't he be in Bolivia?"

"No. He's just got a five-year contract with Cosmo."

"I see. A resident director who might even direct you. And a friend."

"Bill, for it to be anything more than that, you and I would have to have come to some pretty bitter and final decisions."

"You mean you would, Betty. I made up my mind quite some time ago."

She looked at me, as the late afternoon wind blew in from the sea. It was suddenly cold, but bracing. We were on a beach again, and a big towel was draped around her bare shoulders. As the wind blew she shivered, but more from nerves than from cold – as if it was a really big and difficult scene and she wasn't sure what was expected of her.

"You're saying that nothing, absolutely nothing, would ever change things for you?" she said.

"No. I'm saying that we're married. For my part, I know why. I'll hang in and see how things work out. I won't know if I don't."

"That makes me sound pretty shallow."

"You think you have a problem," I said. "I would think a problem should be given at least a few years to declare itself properly – to let us see whether it's chronic, or acute and self-curing. I think, for what it's worth, we have such a good thing going that we ought to strive with all we've got to keep it."

She sighed. "It's true, what you say, when we're together enough. It's when I'm alone and tired, and you're a world away that I begin to feel . . . well, just overwhelmed."

"Lots of people – let's say millions – were separated for years in the war, and somehow their marriages survived."

"But that was forced on them by circumstances beyond their control."

It reached a point when I said, "See, Betty, it feels to me you're almost threatening me. Saying that if we – you really mean me – can't do something to fix things, you can't be responsible for what happens. And by my doing something, you can only mean that I should get a job near where you are doing most of your work. Well that would be okay, except that you aren't in one place all the time yourself, and the job I want to do for some time to come means I have to exercise some mobility.

"You know," I said, "if there could be anything that could start to 'whiteant' our marriage it could be your pressuring me, instead of waiting to see what I'll actually do myself. Can't we be patient? If we can't, is the thing worth as much as we say it is?"

It ended there. She went back to Hollywood to finish the Bolivar film looking more or less herself again. Happier for the moment, anyway. Each of us felt filled with the other, sated, our love intact . . . sort of.

I felt stronger than before she came, but I knew I might need all of that strength, and a good deal more, in the months and years ahead.

Chapter 45

Half a year remained before I would be returning to the Barrier Reef with Andrew Peacock. Now came my other responsibilities in Australia. I needed to establish contacts with researchers investigating the ecology and conservation of populations of vertebrates – kangaroos, emus, echidnas, platypus, ducks, magpies, freshwater and marine fish. I needed to learn from them what they were discovering about Australia's unique vertebrates, and of the roles these animals played in the ecosystems of which they were a part. I knew some of their work had attracted international attention in recent years, but had a feeling that a good deal of this attention came from people's fascination with animals that seemed strange and new rather than from the intrinsic quality of their findings. As my knowledge of the work increased, I realised the unlikelihood of a satisfactory answer to this problem.

What complicated my task was that Australian biologists, in investigating a previously unknown fauna, were inevitably making "discoveries". In fact, a few of them hinted, only half-jocularly, that a measure of fame should be a routine outcome for any hard-working Australian biologist. I thought this was nonsense: for a discovery to be famous, it had to be a non-trivial discovery. There would certainly be some discoveries the like of which had never been seen before.

But how many of these things would significantly enlarge one's view of the biological world?

Most of the researchers I approached were either university faculty members or employees of state and federal organizations involved in science. Nearly all were happy to discuss their work in detail, and agreed to write comprehensive accounts of it for a book.

My impression was that wildlife researchers were doing much of the ground-breaking and imaginative work in Australian biology that was arousing interest in America, Britain and Europe.

Studies on red kangaroos had demonstrated adaptive responses to extreme environmental heat and aridity, including a mechanism whereby the females could retain fertilized eggs alive in the uterus for prolonged periods, if feeding conditions were too poor to ensure survival of pouch young. Then, when food supplies improved, normal uterine development of embryos could occur, and after birth the mother would produce ample milk and the young could be successfully reared in the pouch.

Other work showed that the hopping mode of movement in kangaroos, which had long been regarded as an almost inexplicable biological anomaly, actually resulted in energy being stored during hopping in the long, elastic tendons of the hind legs. This allowed for a more economical way of covering long distances at moderate speeds than the normal four-footed way of the placental mammals of other regions. This, then, was another remarkable adaptation of animal life to the extremely dry, and often food-poor, conditions of some of Australia's largest terrestrial ecosystems.

The Mallee Fowl, a goose-sized bird living in dry, sandy, scrubby country, was known to make a huge nest, a mound of soil and vegetation, in which to incubate its few eggs, and to spend almost the entire period of egg development, alternately excavating and filling in the mound. It had been determined that the bird's apparently frantic activities – constant opening and closing of the egg mound – were directed at maintaining the eggs at a constant temperature. The source of heat in the mound was decaying vegetation mixed in the sand. The Mallee Fowl's beak was the temperature sensor that informed the bird when temperature in the proximity of the eggs was approaching an upper or lower limit for continuous incubation. The astonishing, comic-heroic activities of the birds on the egg mound was explained by this study.

The Mallee Fowl's behaviour is the sort of phenomenon that

creationists delight in: complex and odd enough to attract a lot of attention, focussed and precise enough in execution and outcome to persuade them that the hand of a benevolent deity must be directing the activity of so simple a creature in an intricate task ... whose end result was a successful hatching! By contrast, for those of us with interests in the fundamentals of evolutionary processes, the challenge was different – that we would eventually chart the tortuous evolutionary history of adaptive responses, caused by natural selection, that had resulted in such bizarre behaviour.

*   *   *

The last example of research I will mention had a different sort of beginning. Australian farmers of many kinds had always suffered from the threat of severe water shortages. Moreover, it had long been believed that many land areas that were too dry to farm could be rendered productive if only a water supply could be brought to them. To many people it seemed obvious that by constructing huge storage dams on the waterways of the great inland river systems of the states of New South Wales and Victoria for a graduated release of water to areas of farming potential on the dry plains below there could be huge economic benefit. The dams would also serve another purpose. Although these big rivers usually did not contain a great deal of water, and indeed could be reduced to trickles or pools in drought years, their watersheds were huge. As a consequence, in those years when heavy rains fell, the flood plains of the rivers could be inundated over quite enormous areas – often leading to devastation and loss of life among those communities who were actually managing to farm successfully on the normally arid plains. So the dream was that the dams would supply irrigation water for fruit and vegetable farming in otherwise dry soil, but would also protect the sheep, cattle and wheat areas from the devastating effects of great floods.

But if the dams proved successful in their essential aims – and this has long since been a hotly debated point – there was at least one criticism aimed at them almost from the outset. This criticism had to do with several species of large fish native to these rivers that have evolved in comparatively recent time, paleontologically speaking, from perch-like, marine ancestors. For about a century or more it had been claimed that there was something remarkable about the breeding biology of these fish, that was linked in some special way to the flooding of the large rivers they inhabited, and in which they had evolved. Many people believed floods, in some fashion, triggered spawning by the three largest and most common of the perch-like fish – the Murray Cod; the golden perch (sometimes given the unlovely name of yellowbelly); and the silver perch. Of these, the Murray Cod was the glamour fish, an animal capable of reaching a great size – the record was about 120 kilograms – and with esteemed eating qualities.

When the big Australian rivers overflow, the flooding can take a long time to dissipate because so much of inland Australia is of very low relief. Planning with this knowledge in mind, John Lake, a biologist of New South Wales Fisheries, constructed two types of pond at an inland fisheries station at Narrandera near the Murrumbidgee River in western New South Wales. One of his pond types was a simple, sloping-sided bowl-shape. The other type was rectangular with an extensive flat bottom and a wide deep ditch along one side. Lake called this type a "flood pond". Both types could be rapidly drained, or filled from an elevated storage pond containing water pumped up from the river.

John Lake had noted river temperatures during floods in the Murrumbidgee at times during the approach of summer when these three species were in breeding condition and when there was evidence that spawning and the production of young had occurred. At appropriate times he filled the bowl-shaped pond to about fifty percent capacity and, in the flood pond, filled the deep ditch. He then trapped mature male and female fish that were approaching breeding condition and placed them in the ponds. He examined the fish frequently to determine by inspection how close they were to spawning condition. When they appeared ready he adopted several procedures. The first of these was to leave the ponds alone. Under these conditions, the fish, in either type of pond, failed to spawn, and the females gradually resorbed their eggs. In the second procedure water was let into the ponds to "flood" them. In the bowl-shaped pond this meant a rapid rise in pond level so that previously dry pond bottom-soil was inundated. A limited amount of spawning then occurred in some instances but only when a particular water temperature had been reached. In the rectangular flood pond, with a very large area of previously dry pond now inundated, highly successful spawning would occur, but again only at a particular temperature that was different for each species.

With a final insight, John Lake performed two other kinds of experiment in the bowl-shaped pond. When the fish were ready to spawn, he let in water of the right temperature but also opened the pond drain so that the water level remained constant. The fish failed to spawn. He then admitted water, not from the reservoir but from the recently filled flood pond. Spawning occurred.

In this simple, brilliant series of experiments, John Lake had demonstrated that the trigger for spawning was indeed flood conditions but qualified by the requirement for a specific temperature for each species. He had also demonstrated that the more closely his experimental ponds could simulate a natural flood, the more successful and abundant the spawning. Lastly, he had shown that mere passage of water through a pond was not a trigger but that flood water from a previously dry pond could be. Therefore, he posited a third factor to go with flood and temperature – some "quality" of the water, acquired as a result of its flooding over a large area of previously dry earth. In the years that followed, Lake performed various experiments in a search for this "essence of flood", but without success.

The particular ecological logic of these spawning triggers – a substantial flood and the appropriate early summer temperatures – can be best understood in terms of adaptive evolution when it is realised that another fish species, the tiny firetail gudgeon, also reproduces under the same conditions as the large species, and that in the extensive flooded areas of land adjacent to the rivers, huge populations of plankton organisms quickly appear. In other words, a flood equals abundant food of gudgeon and plankton for the young of the large species. The reproductive biology of these large fish has evolved to be in lockstep with the occurrence of floods because, in this land of sporadic rains, this is how the food supply to their growing young can be ensured.

Lake's experiments explained what had previously been blocks to people's thinking about the way floods operated as triggers to spawning. It had long been pretty obvious that extensive spawning did not appear to accompany brief or insubstantial flooding. This was reflected when the water rose in the bowl-shaped pond. On the other hand, given the right temperature, there had been massive spawning in the flood pond which simulated a large area of earth covered by water during a natural flood. Now it could be seen that these fish had the potential to spawn every year, but in some years rains came too late, or were too meagre to cause a flood. Then the fish absorbed the products of their gonads and waited for another year.

Above all, the results showed clearly that the storage dams constructed on the waterways of the great inland river systems, by greatly reducing the incidence and extent of floods, severely affected the frequency and success of reproduction among these fish species. This information explained why there had been such drastic declines in their numbers since the turn of the century. It also held important lessons on the need for stringent concern for habitat preservation in all large engineering projects aimed at modifying Australia's meagre supply of inland waterways.

Chapter 46

Those were examples of real science, breakthroughs that Australian scientists could be proud of, which I became familiar with during the two winters when I met with their producers and persuaded them to write accounts of their work for the book Rudy and I would edit. Among the essays, I included one by Andrew Peacock. It would not be of the calibre of the best of the other chapters, because the turtle work had not yet progressed far enough to have been published in major scientific journals, but even in its present preliminary form I believed it had enough merit to be among the essays.

I laboured again with Peacock in my second summer, though the work was not as gripping because most of it was more of what we had been doing the season before. I recalled ruefully that the need for repetitiveness in data collection had been what Betty long ago declared uninteresting and dull at Hellhole Springs. There seemed to be no escape from it – as I had told her then.

* * *

Betty and I exchanged a few letters, but now the feeling came to me that things were really coming unstuck. Where she had formerly sounded pessimistic and depressed, her letters had become more cheerful and buoyant but also more difficult for me to relate to. I got an overwhelming feeling that her loneliness was being relieved, and it seemed clear what that meant. I still loved her and assumed I always would. We had not quarrelled, nor grown to dislike each other's ways, and it had never occurred to me that the reasons why we had loved in the first place might fail to seem as important because we were apart. What I felt was numbness. As I saw it we were, for no sufficient reason except her sense of alienation, going to lose everything we had. I felt anger, the anger of emptiness at events and circumstances that didn't seem of crucial significance yet had somehow assumed a density we were incapable of setting aside. My God! I thought, things can end with neither a bang nor a whimper, but just with a sort of creeping depletion that you couldn't even put a name to.

* * *

At the beginning of my second winter in Sydney, on one brisk April morning, Percy Swanson came into the office I had been loaned and, in his usual abrupt way, asked me if I would come to see him at his office in half an hour. I said yes, and he left without another word.

When I entered his office he looked up from his desk and gave what I realised after a moment's confusion was his version of a friendly smile. "Ah, yes," he said, "what I wanted to ask you, Bill, is whether you had any intention of staying on here after June. I mean will you have completed this task you're engaged in by that time?"

I didn't like the sound of this. He had, it was true, been hospitable enough to let me use the small office I occupied. But now I wondered if he was about to ease me out.

"Well," I said, hesitantly, "I had planned to finish up about next October, but if that's – "

"No, no. It's not that. I just wondered if you'd have any interest in a job here."

"A job?"

"Yes. We're about to fill a special senior academic position. In fact, the Department's recently received a major endowment, courtesy of the widow of Professor James Hilbert."

"Hilbert?"

"Yes. Did you know him?"

"Yes. I was here as an undergraduate in his last years as department head."

"Of course. I'm forgetting. You were in classes of mine the first year Hilbert appointed me – his last year before retirement.

"Anyway, Mrs Hilbert's gift is large enough to endow a special research professorship. If that would be of interest to you."

I suppose I caught my breath or looked astonished or something, but he went on quickly, "Of course, I realise this might not be as attractive as travelling around the world and influencing quite big things by your work and advice, and that in America – "

"No," I said, extending my hand, palm towards him. "The point is that I am tremendously surprised that you would even think of me for this position."

Swanson looked mystified. "Why?"

"Well, my age, my reputation – or rather, my lack of them."

"I see. Perhaps you ought to have the humility to leave others to decide such questions." Again, he almost smiled. "If you're perhaps fishing for compliments from me, don't bother. I try to deal strictly with facts when it comes to judging other people professionally. And I don't joke around with serious matters. You can take it as read that I wouldn't mention this to you if I didn't regard you seriously as a candidate. Besides," he went on, now exhibiting a definite thin smile, "I'm not offering you the position quite yet. All I can tell you is that it is a full research professorship, with only minimal teaching requirements, that the appointee will be free to do any kind of zoological research he wants, and that the appointment will be by invitation. All that is required of possible people for this is that they are asked to submit a c.v."

I didn't answer for some time, and could see, from the beginning of a glint in Swanson's eyes, that he was about to become impatient. I had to make some response.

"It's pretty staggering," I said. "But I have to think about it"

"Obviously," said Swanson, shortly. "It's your decision. However, if you have any interest, you ought to put in your c.v. Then, if you get on the short list, you can worry about making up your mind."

"Will there be interviews, then, and if so when?"

"There will be, eventually. But not many. Up to six people are being invited to submit. One will be chosen. In three months."

He looked at me evenly. "The opportunity could be wonderful," he said, "but only if it genuinely appeals to you."

I swallowed hard. "I know," I said.

For two days I kicked the thing round inside my head. By that time the inside of my head was hurting me. I submitted my c.v. to Swanson, though . . .

Chapter 47

I carried the single sheet of paper to Swanson's office holding it flat and open, with the creases of where it had been folded keeping it stiff enough not to hang limply from my fingers. His secretary said to go right in and I went.

He was actually smiling openly. "Ah," he said, "you've got the invitation, then. Good. Have you decided?"

The time for mincing matters – something I have never been good at, anyway – was over. "Look," I said, "this is all very well, and I know I ought to feel great, and in a way I do, but I have to ask you some things."

"Go ahead," he said, though the smile had got cold.

"Here goes, then. Ever since I was an undergraduate in your classes I have admired your mind but also felt that our ideas about biology were apt to be in permanent conflict. Now I'm offered a full professorship in this department, a rare appointment in an Australian university for other than the chairman. But though you're head of this department, if I accept this, I won't be outranked academically, even by you. How is that going to be for you? And for me?"

"I suppose what you mean is that I am what is called a reductionist," he said, " and while that is okay for a cell biologist or a

physiologist it's not considered – according to views currently popular
– appropriate for an ecologist."

"It's not that I care about your being a reductionist," I said.
"I'm often one myself. It's that with me reductionism is part of the
scientific pragmatism I adopt in trying to crack problems. With you
– and here I really want to avoid offending you – it seems more like
a polemical attitude. . . or at the least an overly rigid philosophy."

He surprised me by laughing. "You mean that I stick to it in
an unreasoning, stubborn way? That it clouds my judgement?"

"Not necessarily," I said. "I think you have one of the
clearest, most cutting minds I've known. But I think your particular
attitude may sometimes militate against opportunities and solution of
problems in research. At other times, no one I've known sees more
clearly."

He looked at me for a long time. "Logan," he said, " you
may as well know that my personal support of you for this appointment
comes in three ways. First, I read a letter of reference for you that
was solicited from Rudy Hauser, some of whose work and thought,
despite obvious differences of approach from mine, I can admire.
Second, I thought your work on the thermal spring, on the coyotes
and on the zinc-polluted river – all of which I have read – were
ecologically sound and of much scientific interest. Lastly, though my
approach to the philosophy and the concepts of biology are somewhat
different from yours, I read your Gustafson Lecture, which I found
original, enjoyable and . . . inspiring. Of course I didn't agree with
every aspect of it . . . Anyway, in my opinion, and that of several
people who expressed their views as to your suitability for this position,
you've demonstrated that you have unusual ability as a scientist. Your
relative youth just says to most of us that you will have plenty of time
to do work of even greater merit than you've managed so far." He
smiled again. "I feel that our differences need not set us at odds. I
would like to have you here. And I would like to welcome you."

Slowly, as if fearing a rebuff, he extended his hand.

As what I was about to do fully dawned on me, I reached
out and shook it.

*   *   *

I wondered what other Australian biologists would think of
my appointment. Would I be regarded as an upstart? Worse, as a

renegade, who had gone overseas to get a reputation, and who would now outrank those of more solid, homespun virtue who were toughing it out in Australia? Well, I supposed I would just have to accept whatever came.

Meantime, I cabled Rudy Hauser. I had written him three months earlier to let him know I was going to allow my name to go forward, and he had replied encouragingly, saying that he "figured something like this would happen before long.

"I don't want to lose you from IEC, Bill, but I don't see why that should happen, because you can still remain on the 'faculty' of the IEC, and as a research professor you'll have a lot more clout to do the sorts of things you've been doing already if you want to continue in that direction, as I sincerely hope you will.

"Just one thing troubles me about all this: what of Betty and you? Forgive me for prying, but you are my friend and I also think that gal of yours is a rare creature. By chance, I met a fellow who was friend of my youth, and who now works in movies. I was telling – I guess in a boasting kind of way – that a young colleague of mine was married to this remarkable young woman and he said: 'My God, Rudy, don't you know that marriage is supposed to be on the rocks? Betty McMurtry, according to the gossip columnists, is involved with some Italian director or producer, or something.' Bill, maybe I should just shut up about this. Tell me, if I should. But God, I hope this is all just a crock. It'll be too bad if you two can't work out a life."

I wrote now to Betty. I had not heard from her for more than four months, though I had sent her two letters from the Barrier Reef, to be conveyed by the weekly launch to the Queensland mainland, from there by plane to Brisbane, and then on by airmail across the Pacific to Hollywood.

I told her the offer I'd written about had come through and that I had accepted it. I explained that because of the nature of the appointment with its meagre teaching duties and my continuation as a member of the IEC, I would remain mobile, moving around the world on various research assignments. I wrote, "The appointment itself will make a major difference to the kind of professional life I've been pursuing only in providing me with an academic status that will give me more influence in getting my work done. I dare say we'll have no less chance of being together at least part of the time over the next several years than has been the case during the last two.

That is – and here I hesitate even to say it, but feel I have to, because it's so long since I heard from you – if you still want that."

I also wrote to Andrew Peacock. "There's little chance we'll be able to work directly together again, at least in the near future. But I'll remain interested in all you produce on the turtles and, after I get to know the political and economic possibilities of Australian science a bit better than I do now, it's possible I may be able to help you get your studies better funded. Of course, I may be able to do no such thing. Let's stay in touch, though."

Andrew Peacock wrote back in characteristic vein: "Marvellous news, Bill. Good to know someone got this who won't allow himself to become involved in too much of the bullshit that seems to go with high academic rank in this country, a form of bullshit that seems much less acceptable here, in the raw Antipodes! than in a tradition-laden dump like the U.K.

"It's been good to work with you. I only hope you won't waste yourself doing things that various useless bastards will expect of a 'Professor' who occupies an endowed chair and that you'll feel morally bound to do."

*   *   *

A long month after I had written to Betty, a reply came. It said a number of things, all of them heartfelt, some of them confused and confusing. But it ended by saying, "Bill, whatever happens, I want to say that you have been my only love, and I don't know whether I'm ever going to be able to settle to a life of work, even of work I like, whose price comes so high."

I folded the letter slowly and put it back in its envelope. If that sentence had begun with her name instead of mine, I could have written it myself.

Chapter 48

Following a lecture I had been talking for an hour with a student, Toni Felton, when she became exasperated with me, crying, "All you academics are the same! You know a lot, but you can't get your act together. You don't get your knowledge to the people in Main Street, or the town square . . . or the guys in the pub. You ought to be down there preaching the message."

"Look, Toni", I said, "I'm a scientist, not a politician, not a preacher. *You* go down to the pub and preach, my dear. I'm bloody sure you know the folks there fifty times better than I do. But when you do go and preach, just remember it'll be dry, detached guys like me that rang the bells and wrote the sermon."

Colour flared in her pale cheeks and her blue eyes flashed. "I know that!" She was nearly shouting at me. "That's just the point. It's precisely because you people do know it all, because you have done the work, that you have the authority and the. . . the moral obligation" – the phrase must have pleased her because she repeated it – "the moral obligation, the *duty*, as well as the scientific understanding, to tell it like it ought to be told."

I responded quietly to this. I wanted to calm her down, and didn't want to become heated myself. "Toni," I said, "you'll find out as you live day to day what you should be doing in the world, what you're cut out for and what you're not. Me, I've done some speechifying about conservation in my time, which may or may not have had the slightest effect. But I'm not a mob orator. If someone – someone else – can carry the message of conservation, if a message about it can ever really grab people, I mean *really* grab them, and that someone has the guts and drive to get up in public and spout about it, using a text people like me write for them, then great! More power to them. They'll be better than any hundred bloody politicians, no matter how well-meaning, could ever be. I mean if they really have the gift of the gab that it'll take. But it's not going to be me. I have no knack for that sort of caper. My job's to keep a cool head and keep working and writing so I can feed the true facts and data to the ones that have."

This was in 1967; differences of approach aside, Toni and I were among the many who could still share a basic optimism, that made "how to save the planet" a rousing and hopeful clarion call. There was a buoyancy among all the scientific types – the biologists, geographers and hydrographers who would soon begin to call themselves conservationists or environmentalists. Many of the public, some politicians, even some industrialists were evidently with us. The courts were starting to give weight to our views. Some of us were already teaching relevant courses in the universities, and money was being found to establish departments of environmental sciences and environmental law.

But if I'd only been able to glimpse the true, fearsome lineaments of the future, I'd have taken Toni's admonition to heart. I'd have gone down to the pub and – maybe fortified by a beer or two – spouted. I'd have stood in the town square and delivered an address. No ... Probably I'd have done none of those things. But I like to think I'd have done more than I then deemed necessary – more than I then believed fell within the scope of my abilities.

As it turned out, I didn't do much at all beyond my usual science, and writing about it, and speaking to like-minded scientists. Because, well, it was only 1967. If that could be any excuse.

* * *

By now I had been at the university for a year as a research professor, and had also been an IEC consultant in countries as far away from Australia as Egypt, central Africa and South America. I had also delivered a course of ten lectures on ecological and conservation themes to graduate students in several disciplines. Toni had been a student in one of these courses.

I had defended myself when she criticized me for not having become a popular preacher on environmental problems, and felt right in my belief that the cobbler should stick to his last. My abilities, such as they were, lay in the area of personal action, attempts to solve problems by the use of biological science. I did not believe I would be effective in a public role. And yet I felt depressed, later, about this very point, because I already had a growing sense that though the causes of conservation are good and right and must be defended, they cannot win the day in the world as it is. Because, as a character in the play *Marat/Sade* says, "Man is a mad animal!" Yes, that must be levelled against our species because, as the years grind past, it is impossible to view the wanton acts by so many of its members against each other, and against their environment, as anything less than insanity.

* * *

A day or two before Toni had lost patience with me, I had been eating lunch in the Union Bevery with Percy Swanson. I can't say we had ever become friends. Swanson was, polemically and philosophically, always "in play", you couldn't have a casual or friendly conversation with him, and I found him too didactic, and ultimately too exhausting, to spend much spare time with him. But we usually managed to be civil to one another. On this day we met by accident as we were leaving the Zoology Department for lunch, and now he was complaining to me that his administrative load had tripled in the last few years.

"When I was offered the Zoology Chair," he said, "I was urged to give research leadership to the Department – by example. And I wanted to. But now, my research is negligible and I don't see how I'll ever get back to it." He stared at me in that kind of half-hostile way that made him so hard to feel at ease with. "I envy you. How I envy you!" I shrugged. It wasn't easy to know how to answer him. Then he said: "But, you know – and I have to tell you this, I feel disappointed about the way many of your activities have gone."

I stared back at him. I began to reply, but his next remarks cut me off.

"When I and others supported you for your professorship, it was on the quality of your scientific work and its promise. You've gradually shifted the emphasis so that you're now deeply into conservation – I mean its principles and philosophy. It seems to me that's almost a betrayal of your science as such."

"Come off it, Percy!" I exploded. "Circumstances alter cases. Jesus! you've just confessed that your own work has shifted completely since your were appointed to the Chair. As for me, I was already involved in conservation matters."

He blinked. I saw a look in his eyes that suggested one thing: enjoyment. Nothing Percy Swanson enjoyed – "enjoyed" is probably too weak a word – more than a vigorous polemical debate. I knew I was in for one. I didn't like the idea of this over lunch. It's not the right time of the day for me. I knew that if he persisted I would not digest my meal properly, and I would get irritated and then cross. Which I didn't want. However, it's also a fact that I am at my sharpest when I am angry.

"The point is," he said, "that the increase in my administrative work load is something I'm utterly unable to control. You, on the other hand, can go in exactly the direction you wish."

"So," I snapped, "what's the problem? That's just what I'm doing – going in what I perceive as a necessary direction."

"The problem, as I see it, is that though you may be working hard, you're not doing enough science."

"Look," I growled, "conservation requires hard thought, consideration of a multitude of issues. That's the sort of thing that's taking up my time at the moment. And also, I have to travel long distances to familiarize myself with what's actually happening in environmental hotspots, or to see up close the work people are doing on animal populations."

"Why don't you get hold of some particular conservation problems that call for real science by you to solve them?"

"Such as?"

"Such as the work on coyotes that you did in the U.S., or that work on the polluted river system?"

"Sure," I said, "that's the sort of thing I like best, but many problems that call for conservation measures are a lot more multifaceted than that. You may have to consider socio-economic issues. That's what was wanted in Indonesia with the tambaks, and it's damned difficult. And it's what's happening with Andrew

Peacock's work. He knows what has to be done ecologically to conserve green turtles, mainly because of his studies of their population biology on the Barrier Reef. But to translate that stuff into a plan for the global conservation of the green turtle calls for some understanding of the economic and social forces that are going to lead to their eventual demise."

"Okay," said Swanson, "but you ought to leave those issues to others – economists, politicians and the like."

"That's bullshit, Percy," I snapped. "It's largely because of stupid economists and politicians that the world's in the bloody great mess it's in today. And getting worse by the hour."

"I see. And you, who are supposed to be a biologist, understand economic theory and politics better than – "

"Look. The economists have invented systems so complex and often so perverse that no two of them can agree. But most of what they say is based on total misconceptions about the scientific nature of the human species and its relation to world resources."

"I still don't see – "

And so it went. The trouble was, in a secret sort of way I felt much as Percy did. I was getting further away from the science I really wanted to do. I had in fact already moved part way towards what Toni had wanted, but only part way: I still avoided 'preaching' in the sort of public, populist, political way she had demanded.

Next month I was supposed to be looking at the problems of environmentally endangered species of cliff-dwelling seabirds in Europe at a conference in the south of England. I would not be doing any fundamental science, only trying to help evaluate the significance of recent research, so that conservationists could come up with an international plan to protect these seabirds – a compromise, in all likelihood. And that was another thing that grated: the usual outcome of one's recommendations was a compromise, hardly ever a clean, direct solution based on the known facts. And finally I was being used as a consultant rather than an actual scientist.

*   *   *

I had dined alone . . . which was often the case . . . and because it was a fine, warm evening, clear and moonless, I went down to Circular Quay and caught a ferry for the village of Manly across Sydney Harbour, which is just inside North Head. The

Harbour shores near the city were crowded with lights, but it grew darker as we pulled away towards Manly and, as we got into the Harbour where it was widest and furthest from any shore, I could begin to see the stars from where I stood on the open deck in front of the wheelhouse. The warm breeze pulsed on my face and all my being went back to the first night on the Sausalito ferry. Only now, what I felt was not elation, joy, the anticipation of what would be the best thing of my life, but emptiness, something edging towards despair, and a longing to know where she was and with whom . . . and what she was thinking.

That day, I had received a letter from her; the first in a long time.

The letter said a number of things in a fairly superficial way that we both would have scoffed at three years earlier. Now I read them with as much absorption as I had ever read anything. But the effort was vain. If they were a screen to conceal real meanings they worked splendidly. I did not believe that was their intended function. I read on until I came at last to the part of the letter for which the rest of it was just an impact cushion: "So, Bill, my dear, I think it is not reasonable for either of us to continue in this way. Nothing appears to have changed in our work situations, and we are both, in a sense, in chancery, hostages to each other's inability to reconcile our ways of living so that we can be together. I believe we should separate legally. If that seems a step that neither of us ever expected to face, I now believe it is the only way we can get on with the lives we have without the perpetual concern of each of us for the effects of our actions. I hope you agree. Please write me as soon as you can. I want to proceed on this matter. Love, Betty."

If the letter had been shorter I might have found myself chanting it like a mantra. As it was it thumped in my skull like a high velocity bullet that has penetrated a tank and is ricocheting around creating havoc and merry hell among the creatures inside.

Chapter 49

Somehow the night passed and with the dawn I began to write a reply. I fought down my basic inclination to write a long, beseeching letter that would examine what had happened to us, and concentrated on trying to get at the core of things. I wrote:–

To put it as simply as possible, Betty, I never thought of you as being in any degree a typical modern American woman until now. You were just the girl I loved and married. Nor did I much think of you as the film star, the iconic beauty that thousands dote on, the critically acclaimed actress who is also in the process of amassing a movie fortune. I believed that, in all the ways that mattered to me, you were as untouched by the huge public relations and market forces that surround you as anyone could be. At your centre you were just as you had been, so I could continue to love you, and to count on your love. I know you have felt great loneliness in recent months, as I have too. But I thought that we are both still young and it might well turn out that before too long either or both of us would become less mobile in our work, eventually making it possible for us to be together more.

Anyway, all this is only relevant because now I do seem

to have some reason for thinking of you as a child of your society – a victim, if you like, of its current habits, since now you want to resort to divorce as the solution to our problem, a problem not of being unable to get along, but only of not being together as much as we both would wish.

I can only say, for God's sake let us hang on, at least for another year or two, and meanwhile get together as much as we possibly can to see what we can make of all this.

I was on the Harbour last night. I went there because I was melancholy, and to think of your letter, and because I wanted remoteness and peace – so far as they can be had in a large 20th century city. I did think hard and long out there on the black water, but I thought more about us on the Sausalito ferry than about anything else. And then, Betty, I thought about when we took each other on, when I believed it was as much about putting up with whatever came our way as about continuing to love each other. Let's remember that. This is not something to get divorced over. Because that won't solve a damned thing. Not if we still mean anything to each other. Don't start to talk and act so American, my dear. I love only you.

*   *   *

But the die was cast, or something . . . As soon as a reply could be expected I got it. She said she understood what I had said, but another person was now involved, and that it would not be fair to him if we were to continue like this.

I thought about this and all it implied in its brief text that said so little and so much about what she must think of what I had said or written. And I realised at last that, with ridiculous self-confidence, I had much misread the one person in the world whom I had believed I truly understood. I thought I would never again feel sure in my judgement.

I also felt a clinical coldness; more than that – a deep chill – as if I had lost the best thing that was ever likely to be mine. At first I fell into that simplest and stupidest of parlour games where you say to yourself: I loved her more than she loved me; I am a victim in a tragedy. But in a couple of days, I realised I had no such knowledge. It could easily have been the other way round. Since I had managed to endure what Betty was apparently unable to endure, it just might mean that her feelings, not mine, were the stronger.

In the end, I took a clean sheet of paper and, starting just above the middle of the page, wrote:

Betty, my dear Wife,
The clearest expression of what I feel for you will be that, regardless of my belief that what you propose will be the worst of all mistakes for us, I will accept it, if you absolutely insist that it happen.
Love, Bill.

Chapter 50

     I cannot say exactly how I endured the next few days, though I suppose the most hoary of all consolations – work – played a major part. Then I received a letter that informed me of an upcoming conference in Sao Paulo. It said that "The subject is to be 'Anthropogenic Influences on Physical Environment and Organisms: A Challenge of Global Dimensions?' The organizers wish to invite you to present a paper treating some aspects of your well-known research on a thermal spring system and on your experience with a river ecosystem polluted by zinc. In each case, we would hope you could relate your specific research experiences to broader background implications. For instance, the thermal spring work could be set in the context of environmental warming from such influences as the effluent from industrial operations and power plants, as such influences shape ecosystems and alter biotas, and the zinc pollution could be related to the general problem of increasing quantities of heavy metals in the environment. You might wish to present two papers." The letter included an offer to pay my return air fare and accommodation in Sao Paulo.

     An enclosed brochure described the scope of the conference. One of the many things it pointed out was that after World War I climatologists began to record elevations in Northern

Hemisphere temperatures, and as the years passed with no indication of a lowering, some began to wonder if they were observing the start of a long-term phenomenon. However, in the early 1960s the temperatures fell back to something approaching what they had been before 1920, and it seemed possible that what had been observed was just a cyclical climatic fluctuation of a kind that had been previously recorded. But other events had been occurring in parallel with atmospheric warming. They were almost all related to increase in human numbers and the rise of modern industries. Carbon dioxide emissions from burning of fossil fuels and the manufacture of cement – as measured in tonnes of carbon – had risen from about 1.6 billion in 1950 to nearly double that in the 1960s. This potential addition to atmospheric carbon dioxide was not often viewed with concern, it being assumed that it might do no more than increase world plant production. But that result would, of course, depend on whether the already greatly increasing global demand for wood products eventually reduced the volume of world forests to a serious extent, and also whether farming and grazing lands remained constant in area.

There was no doubt whatsoever that the by-products of industry in the form of various metals – many of them toxic – had been pouring into rivers and lakes in ever-increasing quantities since the start of the Industrial Revolution. The great rivers of Europe, such as the Rhine, showed the quantities of zinc, copper, lead and arsenic all undergoing threefold or fourfold increases from 1900 to 1960. Our work on Dixon Creek had just been a microcosmic example of how such industrial metals got into river systems and worked their mischief.

Then there was the phenomenon that would soon have the universal sobriquet of "acid rain" – something full of threat for lakes and their fish, and probably for forests, in every industrialized part of the globe.

There were bound to be complexities and subtleties without number among these phenomena, some of them of less importance than others. But many conservationists were already girding their loins for a struggle to occupy them for the rest of the century and far beyond, trying to figure out strategies and research plans, how to persuade governments and industries to become more responsible for what they were doing.

Yes, I liked the idea of the Sao Paulo conference.

I wondered how, if I went, I could somehow manage to get to California from there. And whether I should try.

I wrote to Rudy Hauser, explaining my interest and asking for comments. He replied at once, saying that he thought it imperative I attend, and offering to cover any expenses of a trip from Sao Paulo to San Francisco to meet with him and discuss what I was doing in Australia.

* * *

As far as the conference was concerned, there was an introductory address in Portuguese given by some high government functionary (Brazil was being ruled by the military) – a long, rambling talk that had little to do with conservation but was full of inflated rhetoric. In the long pauses that punctuated the rolling oratory of the original, it was also given haltingly in English translation by a pale, timid-looking man. Otherwise, the conference can be said to have lived up to one's expectations. I mean that there were a lot of "case studies" – reports of people's actual research on specific examples of the effects of human activities on the environment and its contained plants and animals, plus a number of "position papers" that dealt more with general approaches to conservation, sometimes considering the broad lessons to be drawn from a major example of environmental degradation. The meeting lasted three days and wound up with a six-hour plenary session which was good unruly fun with plenty of denunciations and spirited, impractical calls for the harnessing and mobilization of conservation forces on a global scale. Among the most interesting items at the conference were historical reviews of changes in the Amazon basin that had followed European settlement in the Americas, the spread of the world's deserts in the 20th Century, and warnings of the effects of over-exploitation on the world's major fish stocks.

Perhaps the most important outcome was that many of us were at least able to appreciate that, if global climate did eventually change, the results could have significance for everyone on earth.

* * *

I did somehow find my way to California. I had not let Betty know of my Sao Paulo visit, nor of the possibility that I might make it to Hollywood. I just got a taxi from the airport and rang her doorbell. She answered herself. For a moment she couldn't see me properly as I stood with the huge red ball of the setting sun behind me. "Hello, Betty," I said.

"My God!" she cried, "Bill!" She took a half step backwards and I almost thought she would fall.

I moved towards her. "I'm sorry," I said. "I really should have told you. But I was in Sao Paulo at a conference and I just finally decided I better come up here. And, well . . . I suppose I didn't want you to dissuade me."

She stepped forwards then and kissed me. "It's good to see you, Bill. Come in."

She was still living in an apartment, had never bought a house. But it was much bigger than before, comfortable rather than luxurious, furnished in eclectic but attractive style. We entered a large, sunlit living room, made to seem more opulent than it really was by a large oriental rug and several comfortable leather chairs. Near the window was a man, perhaps four or five years my senior. Without hesitation, Betty said, "Bill, this is Franco Olivetti. We met during the shooting in Bolivia, and became friends. Franco's employed by Cosmo now and will direct a film for them next year."

He was a decent-looking fellow, and his handshake was firm, though he looked pretty disconcerted as we eyed each other.

"Will you be in the film?" I asked, looking across at her.

There was a slight hesitation before she answered. "Yes. The Cosmo writers and Franco are working on the shooting script at present. It'll be based on Bernard Shaw's 'Arms and the Man."

"Really?" I said. "Will that do well at the box office?"

"If the script is good," Franco said. "Certainly Betty can handle a Shavian part."

"I'm absolutely sure she can. Glad to hear it from you, though."

He looked suddenly self-conscious, even anxious. "I must be going, I have an appointment at six thirty. I – "

"Well, what I want to see Betty about won't take long. I just want a private word with her."

"I know the situation between you and Betty – "

"Do you, now?" I said. "That's interesting, because I can't pretend I really know it. And I'm sure as hell in the dark about the situation between *you* and Betty. Maybe you can explain that to me."

"Bill, please . . . " Betty said.

"Please what? I speak the truth. I don't understand the situation. If you know, I'd like to be told in great detail before we part forever. Or if Franco knows, let *him* tell me."

Out of the corner of my eye I saw Franco draw himself up. "Betty is always a very honest person, Bill. She has told me many times that she has loved you – "

"*Has* loved me?"

"Pardon. It is my English. Loves you. But that it has been very difficult for her with you away so much. She – "

"With respect, Betty has also been away a lot, and both of us knew that our marriage would not alter things like that, at least for a number of years." And then, with all my efforts at control starting to come apart, I said, "And, also with respect, Franco, so far as I'm concerned it's none of your damned business. So, as I've only a short time to speak to my wife, why don't you get lost?"

Betty started and I saw Franco's eyes flash – and I got ready.

Then I stopped. If I didn't believe in Betty's answer to our problems, neither did I believe I would ever fight anyone over her. Fight for her, to protect her life, her reputation, to guard her against harm, of course; but not over her. This was my wife, my beloved. If, for whatever reason, she chose another, I might be hurt to my mortal core. But I would never go to war over her. It seemed to me that my love was better than that.

Anyway, Franco saw my point or something, and left quite suddenly, against Betty's apologies. She came back, eyes filled with tears. Before she could speak I put up my hand.

"Look," I said, "you're a wealthy actress. Get on a plane with me and come to Sydney. Live with me there until your next film. It's a year away, isn't it? Let's have what life we can at present without bemoaning our fate. Just come. You know I can't be here long, because I simply can't get the long breaks that you get between films. Come. Now."

"I . . . I want to, Bill, but I mustn't. If I did, my resolve would weaken. It would all be great again for the time we'd be together, then harder than ever."

"Betty, live for the moment: 'Sufficient unto the day is the evil thereof. Take no thought for the morrow'. There you are, two biblical quotations. From a sceptic like me. We live in a world of fears and terrors. Don't plan. Just live, and try and make the time seem eternal. We'll get it all together, someday. Perhaps some day soon. We mustn't suspend our love for reasons that are as unimportant as being separated from time to time."

For an hour we argued. Then I started to feel cold inside. And suddenly it was over.

"As I said in my letter, if you truly want a separation, you can have it."

I kissed her goodbye at her front door and went out into the early twilight.

Chapter 51

I went on to San Francisco the next day to see Rudy Hauser, but didn't call on him immediately. Instead I drifted around the city for a couple of days, revisiting some of the places Betty and I had liked best. But the associations were flat without her. All I did was torment myself with memories of things that could not be repeated. Memories made me bitter and resentful, and angry for feeling that way. Finally, I just made up my mind to see Rudy without further delay.

* * *

Rudy was not in his small, cluttered old office, but in a splendid new one of grand dimensions and lavish furnishings. He sat at a great polished wood desk on a fancy swivel chair with leather upholstery. The floor was newly carpeted, the walls tastefully pale and adorned with original prints and paintings.

He sat at the desk with his back to a picture window that gave a view of a green parkland. He seemed smaller and older than I remembered and he wore an unaccustomed expression of self-consciousness. He had the look of a man who has long been bald and has suddenly begun to wear a flamboyant hairpiece . . . and is waiting for his old friends to rib him or burst out laughing. I reacted more or less in keeping with such expectations, because I felt myself grinning like an ape and saying, "Jesus, Rudy, did your rich uncle die or something?"

Then Rudy, too, was laughing and gesturing almost helplessly at the huge office that surrounded him. "God, Bill, I feel so out of place in this thing. I was sitting in my little rat hole of a place for ten years, and I only had to reach out my arm and I could find any papers and books I needed, and I felt comfortable there and sort of secure. . . and now this."

"So what is 'this?'"

"Oh, well, IEC's been getting much bigger money recently. We've got a budget of more than three million, and the principal funding agencies have been getting anxious about how a professor with major responsibilities to university research and teaching can really do justice to IEC projects as these expand and multiply. And they're right. So they came to the university with the proposal that I be given a year's leave of absence – this has just happened – to assume full-time directorship of IEC. If I don't like it after that, or if the university doesn't want the arrangement to continue, then we'll all take a hard look at things and some final decisions can be made. And this" – he waved at the office – "is just the front they require of a full-time director." He laughed. "I've even got two full-time office staff, one of whom doubles as a receptionist. I'm not sure yet whether they'll prove really useful, or just give me two more people to be responsible for."

"What about your present graduate students?"

"I'm co-supervising them with several faculty colleagues who've been more than generous in their willingness to help me out. Anyway," he went on, clearly wanting to get down to business, "I will now, supposedly, have more time to concentrate on IEC matters. What was the conference like?"

* * *

We ate the last of our sandwiches and drained our coffee cups in Rudy's posh office. He still preferred to eat in his office rather than in the Faculty Club.

I found I was gazing out his great window at the shimmering day and asking myself why I wasn't a farmer or an artist, or a field naturalist. Or almost anyone able to spend all my time in the miraculous outside world. Meanwhile, I could hear words being spoken in a voice that I recognized – almost with a jolt – as my own, "So, for me, the only really big thing coming out of the conference was the implication of a man-made warming of the atmosphere, and the problem sorted itself out into two principal lines of enquiry."

"Which were?"

"First, to prove that such warming is really happening. Second, to compose a reasonable scenario of what resulting world-wide biological adjustments or dislocations would be likely."

Rudy poked around with a pencil on a pad he had been making notes on. "Okay. But I'm not sure how you'll convince enough people of the *a priori* likelihood of this warming as a constant feature of the future so as to be sure you'll get some acceptance of the biological scenario."

"I've thought a bit about this. It's early days, yet, in getting the whole idea of global warming widely considered. Though if it happens, interest and concern will grow with every passing year. But I think a good start could be made by publishing essays by some of the participants at the Sao Paulo conference that would deal with the influence of increasing carbon dioxide levels, and calculating, on the basis of international growth rates of industry, how much greater the carbon loading of the atmosphere is likely to become during the rest of the Century.

"If a persuasive case can be made, we'll have the basis for a model that can be continually elaborated and refined over a long period. And if we conclude that there is a real case for warming, then, even if we can't be sure about its severity, we can infer certain changes in agriculture, movements of the ranges of plants and animals world-wide, and . . . the gradual reduction in the polar ice caps and rising sea levels."

Rudy sighed. "Yeah – quite a litany of woes. You aren't going to become a sort of ecological Jeremiah, I hope?"

I grinned. "And then there's acid rain . . ."

*   *   *

We talked throughout the long afternoon till the sun was low in a clear and cloud-free sky, and Rudy was saying, "We better get out of here. Nancy should be home in half an hour. We can help her prepare dinner, if that's okay with you."

"Sure," I said. "And thanks for the invitation. But let me just finish like this. I want to ask you, as head of IEC, to request me to initiate a long-term study of these two things: global warming of the atmosphere through carbon dioxide increase, and acid rain. Both are results of industrial pollution, both will grow unless determinedly checked, both are world-wide in their influence, and both will be likely to have profound and wide-ranging effects on the biological world.

"It's not that we should necessarily do lots of work on these things ourselves, but we should be encouraging and fostering and helping get them started. Now. If IEC can't be far-seeing and universal in its aims, it's surely falling short of the vision under which you founded it."

Rudy got up and pulled on a jacket. "Okay," he said. "I get your drift. I'll consult with the board of directors and let you know very quickly where we sit. Dinner now. Or soon. First, though, a drink. And how are things with you and Betty?"

*   *   *

The dinner, when it finally came, was very good – broiled salmon steaks and wild rice with a Californian white wine – and then we sat out on the patio on what proved to be an unusually warm evening for the time of the year, and chatted quietly over brandy and coffee with the stars overhead.

The subject of Betty and me was brought up again, this time by Nancy. She was a quiet and serene woman, but she could also be very direct. "How," she asked, "could you have allowed this rift to happen, Bill?"

"Now just a cottonpickin' minute, my dear," said Rudy. "You can't ask that question in quite that way. Betty and Bill both have their careers. Even if Bill's isn't much as far as income or popular fame are concerned, he puts a lot of store by it . . . and so do I. And Betty has her career – which we all know about. They have a problem to figure out, both of them. It isn't Bill's fault, as your question seems to imply."

"Sorry," said Nancy, "I shouldn't have put it like that, of course. But tell us what's happening, Bill. We're concerned for both of you."

At the end, none of us was stirring, just looking at the stars and holding empty coffee cups or brandy glasses.

"I just hate to hear what you've told me, Bill," Nancy said at last. "Your wife – well, I know she's this great film star – but the main point to me is that she seemed like a really nice girl."

"Well," I said, "she is . . ."

"Then don't let it end there." Her voice rose in uncharacteristic urgency, "Do something!"

"For God's sake, Nancy," said Rudy, "what do you expect him to do? Hit her on the head with a club and carry her off to his cave? You're a female with a full career. You should understand the difficulties. Make a creative suggestion or . . . desist."

Chapter 52

Before I returned to Australia I spent time in the United States and Canada, then went to Britain and Scandinavia. My concern was to meet again with some of those who had attended the Sao Paulo conference and were already working on warming and industrial acidification of the atmosphere. I wanted to get short reports on what they were doing and collect these in a book that IEC could publish. Having also secured Rudy's approval, I asked two of the most authoritative of these researchers to write critical, evaluative essays on the literature of these topics that would help tie together the twenty or so individual reports and make a readable book of the collection. We hoped it would provide comparisons for researchers, letting them know what their colleagues elsewhere were doing and thinking about these problems. It also seemed that such a publication would be a useful tool in importuning governments and industry to find the money and resources for preventive and remedial conservation measures. Lastly, it would be a compact benchmark publication against which progress, or lack of it, could be recorded.

I ruffled some feathers because IEC was still a new organization and few people had heard of it. Many assumed it was some kind of governmental organization, and scientists working for universities were sometimes cautious about this until I was able to explain fully that IEC had a very open and non-political mandate. Those working in government labs were often quick to point out that they could not enter into any arrangements with outside bodies without full permission from their department heads. I was able to straighten out most of these problems. But looking back on my discussions and dealings with industrialists, I recall that we were too ready to assume that they must necessarily be our adversaries. I – and others like me – should have gone to them prepared to explain our concerns in more detail, and with as much expectation of a good and sympathetic reception as when we approached our fellow scientists. If many of us whistleblowers had done everything we could to raise the consciousness of industry leaders about conservation, a less antagonistic and franker attitude might have existed between us from the start. Perhaps, as scientists, we lost our best chances of getting an understanding with industry, just as Brecht suggested Galileo and other scientists had with ordinary people at the time of the Renaissance.

* * *

Back in Sydney I was soon at work on my report for IEC when, one morning over coffee, I found myself again in conversation with Percy Swanson.

"You won't like what I've been doing lately any more than you liked the stuff I told you about earlier," I said. And then told him. I was right. He didn't like it, but for somewhat different reasons.

"You mean to say that you're seriously going to postulate the fact and extent of environmental changes on the strength of a few fragments of data and some speculative notions about their interpretation? And that you'd dare to make these things the basis of long-term studies, with people from around the world involved?"

"To put it in a nutshell: yes."

"You're leaving science behind. This is fortune telling, augury . . . reading tea leaves. You're disappointing me."

"Sorry about that, Percy," I said. "But just consider. If we're wrong no one is harmed. If we're right, maybe a lot of people will be grateful to those who tried to urge them to be ready to prevent calamities . . . or at least tried to figure out how to deal with them."

* * *

I thought it ended there, with a hot and apparently unresolvable dispute. But a few days later I opened a letter and found it was from Swanson, handwritten in his angular, angry-looking script:–

Dear Logan,

    This note is by way of a warning. I do not want to be accused of going behind your back, but I am much disturbed by the direction your work has taken. I detect in it some grandiose notions about the responsibilities of scientists in assuming leadership concerning problems that beset human society. This is all very well, and laudable, in essence. But you and Professor Hauser – whom I take to be your mentor – are undertaking projects and responsibilities that lie far outside the scope of worthwhile biological research. In short, I do not believe you are adequately discharging the duties of your position. Though it may seem a fairly extreme step for me to report thus about a colleague, especially one of your academic reputation and standing, and who recently enjoyed my full support, I believe it necessary to offer my opinion to the Hilbert Trust, the endowment of which is responsible for your present appointment. I am unable to anticipate what the response of the Trustees will be, but feel it only fair to warn you of my intention.

Sincerely,

P. Swanson,
Head, Department of Zoology,
University of Sydney.

    I didn't mess about. I went straight to Swanson's office and put the letter on his desk. "You had no need to be confrontational," I said. "You could have told me over a cup of coffee, or lunch."
    He peered at me with a strange expression, almost a grin. "Yes, I could have. But I wanted it to be clear just how seriously I viewed things. Conversations are easily forgotten, but a letter makes a person sit right up."
    "I see," I said. "May I remind you that you said I 'would be

free to do any kind of zoological work' I wanted; your exact words which I carefully memorized."

"Yes, I accept that. But is this zoology? I think not."

"Balls, Percy! The plain fact is that you've got your own restricted view of what the subject's about and you're intolerant of others' views. I always thought so, and age hasn't broadened you."

He was looking at me now with a hard, tight, deathshead grin. "I don't want this quarrel. But if you persist in your folly, I will pursue it through channels."

"Persist in my folly? Christ, Percy, you sound like some hellfire and damnation preacher. You certainly aren't talking like what I think of as a scientist."

He pushed himself back from his desk and stared coldly at me. "Anyhow, that's how I see it."

"You're a smart man, Swanson," I said, "but I always thought you were a narrow, miserable jerk at heart."

Adrenalin powered me out of his office on a wave of elation, but that was dying within minutes, and I knew I was in for a long and bitter dispute that could be resolved in his favour and mean problems for me. Prompt resignation might possibly be my best option. But that would mean a move, and soon. Where to and to what?

Chapter 53

The next day I wrote to Rudy Hauser, telling him what had happened. "It looks very much as though Swanson will make good on his threat," I wrote, "because he's that sort of guy. Whether I can survive an attack of the kind he's capable of mounting I'm not sure. It's not that I feel cowed by Swanson. I can speak out freely on my own behalf and all that. But Australia is still in many ways a hidebound place. And though my academic position – Hilbert Professor of Zoology – carries a certain distinction, Swanson is still my administrative superior and has lots of clout in this university. I can't easily see people supporting me against him if push comes to shove. I wonder if you have any suggestions."

It is not possible to be serious about science and also to be superstitious. But, notwithstanding, misfortunes do seem apt to arrive in pairs or trios, or sometimes in higher multiples.

Or is it just that we notice when they do? One week to the day after my encounter with Percy Swanson I received a letter from Betty, or rather from her lawyers. The divorce grounds she was using were "desertion". I wasn't going to bother with a lawyer. I replied myself, stating that Betty was away from wherever I was as much as I was separated from her. "However," I concluded, "I suppose Mrs Logan is simply saying that I must not want to be near her because I am so often elsewhere. Since I feel happy and privileged to be near her as often as I am able to be I will offer no complaint of my own. I suppose I should be delighted to know that she seeks no maintenance expenses from me! If she is sure she wants a divorce I suppose she will have to have it."

I sent this off and waited for wheels to turn. I guessed it would take a couple of months. I kept picturing her in the company of Franco Olivetti and feeling very low.

Chapter 54

Two weeks after the dispute with Swanson I got a letter from the Hilbert Trustees informing me politely that, as was their custom, they were requesting me to meet them in a month to review the work I had been doing. I had received such earlier letters from the Trustees and had met them over lunch to give brief surveys of my efforts. There had been nothing expressed by them – two senior professors from the Science Faculty of the University, a company director and a lawyer for the Hilbert estate – except polite interest and approval for what I had done. However, the last paragraph of the present letter noted that Professor P. C. Swanson had recently raised some points that might require a response from me in regard to the direction of my work programme. The tone of this passage was gentle enough. It seemed likely that the Trustees would not act quickly or without reflection on anything Swanson had said. Perhaps they knew what a cross-grained scold the man could be!

I had received no reply from Rudy to my earlier letter, so I dashed off a note, briefly recapitulating the first one and suggesting that, as a further clash with Swanson seemed inevitable, and that his complaints, if strong enough and frequent enough, could make my position intolerable if not untenable, I had better start looking into the possibility of a new job. Or failing that, a return to my former position with IEC, if it could still be said to exist. And what could he advise?

After two more weeks I finally got a response from Rudy:–
"Sorry I didn't respond to your earlier note. I have been in Portugal for several weeks. I believe a modern ecologist – which I thought Swanson was – should be at least empathic with conservationist ideals. I'm far from a first rate example of a conservationist myself, because I've still got too many irons in the fire that are just plain ecology. So as Director of IEC I am, in a sense, sailing under false colours. But Swanson is being idiotic to persecute you. You can't have your old job here, but something can be managed." His letter went on –

Merriman has invited me to become Academic Vice-President and I guess – just because of feeling they've been pretty good to me and still being committed to university life and wanting to put something back – I'm going to accept.

There'll be a search soon for a new Director of IEC. Interested? Write immediately (or telephone or cable me) if you are. I won't, of course, be involved at all with the actual selection process, but I think you'd have a good chance. Your youth, energy, ability to get things done, and the international contacts you've already established would weigh in your favour.

Only one thing. IEC has to be run from here. Australia would be just too far from everywhere. Leaving there more or less permanently might not appeal to you. Or, of course, for some very personal reasons it just might!?

Whoever runs IEC will have to travel a bit, though quite a lot of the travel will be in North America, or to main world centres for short visits. Much of the time you'd find yourself, as I have, here in San Francisco.

Incidentally, I enclose a recent item from one of the local papers. It may interest you!

The item was headed *San Diego-Born Actress Calls it Quits*. It went on: *Betty McMurtry, incandescent star of five successful movies has announced her retirement from further film-making. She will be particularly remembered for her role in "Flight of Fancy" and the box-office blockbuster "Bolivar". It was recently reported that she and her husband, an Australian biology professor, had quietly separated, a matter of much regret to their friends. Miss McMurtry was recently rumoured to be romantically involved with an Italian-born Bolivian film director, but the present status of this relationship is uncertain. Friends speculate that it may have come to nothing. Miss McMurtry has a reputation not only as a peerless actress and great beauty, but also as one of the most pleasant and unaffected people in movies. She apparently feels unsuited to the Hollywood world, however, and after a vacation will be returning to her first love – the theatre. She will head a newly formed company of stage performers as manager-director, and will also act. The new company will be known as the SANFRAN PLAYERS, and will concentrate mainly on plays of distinction that lie outside the usual scope of the popular commercial theatre. Beyond this, her plans – including those for her personal future – remain undisclosed.*

*   *   *

I sat down a few hours later to scribble my reply to Rudy Hauser. I wrote a half page, then crumpled the paper and picked up the telephone.

I spoke to the overseas operator. But the first call I made was not to Hauser.

Chapter 55

It is winter. A cold and windy day. We now live in a place high above an ocean beach. I have a supremely efficient, state of the art, high-tech wood stove, whose fire is as warm a spark of comfort to me as it was to those of our species who crouched before smoky ancient fires in cave mouths. To me, as to them, it is holy and magic. I can explain the physical and chemical phenomena that are fire as they could not, but that ability neither diminishes nor increases the fire's comfort and mystery.

I never believed we would come to some ultimate insight into living systems the way modern physicists think they are closing in on a "theory of everything". I loved the biological world as it appeared to me. I found it wonderful enough to want to study and know something of its structures and processes. My aim was to investigate animal populations and their ecosystem settings, and help promote the means for their conservation.

But the world that was has undergone changes so rapid and stupendous that, like many another ecologist, I was sidetracked into two kinds of race. One race was to study populations and ecosystems before they were wiped out or irrevocably damaged. A second race, that rapidly became the dominant one, was to find ways to oppose, even to derail, the madness of unconsidered global change.

But now I conclude my story at the point when my enthusiasm and strength were near their peak – which was also when Betty and I began to achieve the life together that we had hoped for. What came later, after this "happy ending" in our personal lives, will not bear detailed telling – though I'll try to put it in a nutshell. For Betty it was the post-modern theatre of brutality, hostility, alienation and despair that soured the drama and acting she had loved. For me, it was the end of the great feeling of the 1960s that accompanied the feverishly anticipated birth of a popular enlightenment about Nature and its conservation – a birth that produced a healthy-seeming infant but one that soon sickened and perished of many strange and unexpected maladies. Many conservationists grew old before their time in the worry of watching that death in the 1970s and 1980s.

It's true that the burden of conservation – a burden now with the odour of decay and the taste of rusty iron – was taken up again in the 1990s. But fifty percent more people now live on the earth, many of them new to modern technology yet trumpeting their God-given right to repeat and augment the environmental barbarities of the West.

The computers now in use allow us to analyse and display data and model processes in as many ways as our tired wits can devise . . . and there are probably a million more ways that lie beyond out present inspiration. But if knowledge of the world and the expertise to manage its processes have burgeoned, the world itself has shrunk. The old sense of wonder is overarched by a pervasive dread of tomorrow.

The magic of the world is not destroyed by knowledge, as some have believed, but by insane ambitions and the unexamined life. We may yet save this world, but we must move quickly now, and with infinite resolve. Otherwise, the things that have been going on – huge and monstrous forms of factory-farming, ever-increasing threats to forests and fish populations, the effects of pollution on inland waterways and the ocean, environmental warming – may slip at last beyond our grasp forever.

Already there are many who dream of "terraforming" Mars and are laying down detailed plans to do it. No longer will it be the splendid "Red Planet" of our youth, something to study, marvel at, learn from. It will simply have become an erzatz Earth, another place to populate with a new version of humankind. Already the developers are slavering.

One day, we may even reach the stars – which will be one more reason to hope we may yet find ways to come to terms with this Earth, our first, and so far only, home. If we grow wise enough for the stars, our galactic interests may remain detached, scientific, able to elevate our understanding of the Universe and our true role in it. . .

Many former optimists will doubt that we can change so much so soon. But if we fail, I'm sure we'll find some way to tame the stars and dim their ancient fire.

ABOUT THE AUTHOR

Alan Weatherley, born and educated in Sydney, Australia, has received awards in Australia and the USA for research and writings in freshwater biology. During a teaching career of thirty-five years he lectured in physiology, evolution, vertebrate biology and ecology. He was Reader in Zoology at the Australian National University, Professor of Fisheries Biology at the University of Trosmö, and is now Professor Emeritus, University of Toronto. Since retirement, he has turned to painting, and writing fiction. He and his wife, Robena, are active conservationists. They live in New Brunswick.